# THE YUKON RELIEF EXPEDITION

## and the Journal of Carl Johan Sakariassen

# THE YUKON RELIEF EXPEDITION

## and the Journal of Carl Johan Sakariassen


Edited by V. R. Rausch and D. L. Baldwin

with documentation and annotation

original journal translated from the Norwegian by James P. Nelson

University of Alaska Press

Fairbanks, Alaska

# Cataloging-in-Publication

Library of Congress Cataloging-in-Publication Data

Sacarisen, Carl Johan, 1875–1968.
The Yukon Relief Expedition and the journal of Carl Johan Sakariassen / edited by V.R. Rausch and D.L. Baldwin with documentation and annotation ; original journal translated from the Norwegian by James P. Nelson.
p. cm.
Includes bibliographical references (p. ) and index.
ISBN 1-889963-32-1 (cloth : alk. paper) -- ISBN 1-89963-33-X (pbk. : alk. paper)
1. Yukon Relief Expedition (1898) 2. Sacarisen, Carl Johan, 1875-1968--Diaries. 3. Norwegians--Diaries. 4. Yukon Territory--Description and travel. 5. Alaska--Description and travel. 6. Bering Sea--Description and travel. 7. Frontier and pioneer life--Yukon Territory. 8. Gold miners--Health and hygiene--Yukon Territory--Dawson--History--19th century. 9. International relief--History--19th century. 10. Reindeer--Yukon territory--History--19th century. I. Rausch, V. R. (Virginia R.) II. Baldwin, D. L. III. Title.

F1093 .S23 2001
363.8'83'09798609034--dc21

2001034770


International Standard Book Number: cloth, 1-889963-32-1
paper, 1-889963-33-X
Library of Congress Control Number: 2001034770

Printed in the United States of America by Thomson Shore,
This publication was printed on acid-free paper that meets the minimum requirements for the American National Standard for Information Science—Permanence of Paper for Printed Library Materials ANSI Z39.48-1984.

Publication coordination by Pamela Odom, University of Alaska Press.
Production and text design by Deirdre Helfferich, University of Alaska Press.
Cover design by Dixon Jones, Rasmuson Library Graphics, University of Alaska Fairbanks.
Cover drawings: Rock carvings, dated to 3200–2600 B.P.; from Alta, Finnmark, Norway. (Redrawn from several sources by V. R. Rausch.) Reindeer ornament: Figure carved in rock perhaps 8000 years ago; from Alta, Finnmark, Norway. (Redrawn by V. R. Rausch.)

Royalties from the sale of this publication are being donated to the Nature Conservancy.

We remember those who came....

# Contents

| | |
|---|---|
| List of Illustrations | viii |
| Acknowledgements | xii |
| Prologue | xv |
| Introduction | 1 |
| I. Setting and Background | |
| II. The Cruise in Northern Alaska, 1890 | |
| III. The Summer and Autumn of 1897 | |
| IV. The Congressional Debates | |
| V. The Journey to London | |
| VI. London to Norway | |
| VII. Jackson in Finnmark and Devore in England | |
| VIII. Finnmark, January and February 1898 | |
| IX. The Courts of Justice in Christiania | |
| Notes | |
| The Journal of Carl Johan | 63 |
| Journey to Alaska, 1898 | |
| Weekly journal for 1899. Eaton Station and Cape Nome | |
| Annotations - The Journal of Carl Johan | |
| 1898 | |
| 1899 | |
| Epilogue | 233 |
| The Journal: A Brief History | 236 |
| References Cited | 239 |
| Index | 251 |

# Illustrations

**Page xiv:** Carl Johan Sakariassen. 1960.

**Page 5:** Fladstrand, Norway. Leif Boland, grand-nephew of Carl Johan, could not be certain that the old-style structure was that of Carl Johan's parents, but he stated, "I think it is precisely like the house where he [Carl Johan] was born and grew up." The house and all the other buildings in the area were burned in autumn 1944 by soldiers of the German army, which occupied (1940–1945) almost the whole of Norway during the Second World War. (Photograph provided by Leif Boland.)

**Page 7:** Sheldon Jackson. (From an engraving by A. H. Ritchie, N. A.: in Sheldon Jackson's book, *Alaska, and Missions on the North Pacific Coast,* Dodd, Mead and Company, Publishers, New York. 1880.)

**Page 17:** The act enabling the Yukon Relief Expedition: House Resolution 5173. (Photograph: National Archives, General Records of the U.S. Government, RG 11.)

**Page 20:** Warrant No. 1913, Department of the Treasury (24 December 1897): The secretary of war assigned responsibility for these funds (one-third of the congressional grant) to Lieutenant Devore. (Photograph:

National Archives, Accounting Offices of the Department of the Treasury, 18E-4, R-4, B396.)

**Page 23:** Daniel B. Devore: 1885. (Photograph by Pach Brothers, courtesy of the Archives of the United States Military Academy, Special Collections of the Library.)

**Page 31:** The route of the Yukon Relief Expedition. The small rectangles show areas enlarged (maps, pages 39, 95, and 178).

**Page 39:** Route of the *Manitoban* out of Altafjorden, in Norway. The dots coursing to the south of Alta indicate the direction of localities providing reindeer for the expedition.

**Page 73:** Aboard the *Manitoban*, in the harbor of New York, 28 February 1898, members of the expedition wait to go ashore. (Photograph published in *Frank Leslie's Illustrated Newspaper,* 17 March 1898.)

**Page 76:** Men from Finnmark disembarking the reindeer from the *Manitoban* at Jersey City, 28 February 1898. (Photograph published in *Frank Leslie's Illustrated Newspaper,* 17 March 1898.)

**Page 80:** Personnel of the Reindeer Expedition, in Seattle, 1898. Tentative identifications of four: Top row, sixth from left, Jafet Lindeberg; next to him to the right, William Kjellmann. Top row, second from far right, Carl Johan (in military jacket). Second row from top, third from left, Wilhelm Basi. (Photograph by La Roche, published in Sheldon Jackson's report, 1898b.)

**Page 83:** The steam tug *Sea Lion.* (Photograph, courtesy of the University of Washington Libraries, Special Collections, Negative 8417.)

**Page 86:** The *Seminole*, as illustrated (artist's sketch) in the *Seattle Post-Intelligencer,* 15 March 1898.

**Page 88:** Carl Johan's sketch (original and enlargement) of the fishhook of the Tlingit people, at Haines Mission, 30 March 1898.

**Page 95:** Southeastern Alaska: The route of the *Seminole* via Lynn Canal to Haines Mission. The members of the expedition who were ordered to return to Haines Mission departed from their friends near the headwaters of the Klehini River—a locality indicated by the black bar and arrow. The surviving deer were led during the following weeks through Chilkat Pass and towards the Yukon River on a course as shown (approximately) by the broken line to the north of the pass.

**Page 105:** William Kjellmann. (Photograph published in Sheldon Jackson's report, 1897.)

**Page 124:** The steam schooner *Navarro*. More than thirty people of the expedition spent the month of July 1898 aboard the ship en route from Seattle to St. Michael. (Photograph, courtesy of the Puget Sound Maritime Historical Society, the Museum of History and Industry, Seattle.)

**Page 130:** The steam schooner *Del Norte*. (Photograph, courtesy of the University of Washington Libraries, Special Collections, Negative 7723.)

**Page 141:** At Eaton Station, 1898. (Photograph published in Sheldon Jackson's report, 1898b.)

**Page 153:** Eaton Station, 1900. (Photograph published in Sheldon Jackson's report, 1901.)

**Page 178:** The area of Eaton Station and Anvil City (Nome). The site of anchorage (indicated by arrow) of the *Del Norte* on 31 July 1898 was just to the south of Unalakleet. Carl Johan and Johan Petter walked from Eaton Station to Cape Nome in 1899. Locations of villages through which they passed en route are shown.

**Page 191:** Jafet Lindeberg. (Photograph, courtesy of the Alaska State Library Historical Collections, Alaska Purchase Centennial, PCA 20-88, *Discoverers of the Nome Gold Fields.*)

**Page 199:** Sheldon Jackson and others of the Department of the Interior, at Teller Reindeer Station in 1901. From left to right, the adults are: Francis H. Gambell (physician), T. L. Brevig (teacher/missionary), A. Hovick, Mrs. Brevig, Sheldon Jackson, Mrs. Campbell, E. O. Campbell (physician), and A. R. Cheever. (Photograph published in Sheldon Jackson's report, 1903.)

**Page 208:** Carl Johan (left) and Wilhelm Basi, circa 1899. (Photograph by Snodgrass studio, Astoria, Oregon.)

**Page 237:** Carl and Hannah, circa 1940.

# Acknowledgements

Througout the process of publication of this work, we received advice and guidance from Pamela Odom and Deirdre Helfferich, of the University of Alaska Press. We are deeply grateful for their invaluable assistance, their interest, and their warm support.

The information in this account relating to Carl Johan's journal was assembled over several years from individuals, archives, and libraries, as we sought data on the events that determined and were pertinent to the Reindeer Expedition. From those sources we received assistance that not only answered some of our questions, but the interest expressed by many whom we met personally or who wrote in response to inquiries provided an encouragement for which we are sincerely grateful.

We also retraced a part of the route of the *Manitoban*, and that of Carl Johan in 1898 and 1899, in Norway (by DLB) and in Alaska (by VRR). During the visit to Finnmark province, the warm and kind welcome and enthusiastic guidance of the late Leif Boland and Ruth Boland, of Alta, brought new perception and appreciation of the area of Langfjordbotn (there, the southern reach of the fjord, was the family home of Carl Johan), and Altafjorden, and other places, especially Alta, Bossekop, and Hammerfest, notable for the people of the expedition. Anna Gerd Paulsen, of the Alta Museum, provided much assistance, as did Per Rapp of Alta (whose father was a member of the expedition). Øystein Boland and Anna Boland since have helped in many ways. At Unalakleet, Leonard Brown and Mary Brown extended cordial hospitality, and they generously gave of their time, to relate details of the history of Eaton Station.

# Acknowledgements

We worked at the National Archives in Washington, D.C., and Seattle, Washington, the Department of History of the Presbyterian Historical Society in Philadelphia, the J. Porter Shaw National Maritime Library in San Francisco, and the libraries of the University of Washington. Considerate assistance was given at those institutions, and by the following, in written communication: Alaska State Library in Juneau; Aust-Agder-Arkivet at Arendal; Guildhall Library in London; Hedmarksmuseet og Domkirkeodden at Hamar; H. M. Customs and Excise at Salford; Kegoayah Kozga Library in Nome; Maritime History Archive in Newfoundland; Maritime Museum of British Columbia; National Archives of Canada; National Archives of Denmark; National Archives of Iceland; National Maritime Museum in London; Norsk Sjøfartsmuseum at Oslo; Oregon Historical Society; Peabody and Essex Museum at Salem; Public Record Office at Surrey; Scottish Record Office at Edinburgh; Statsarkivet i Oslo; Statsarkivet i Tromsø; Statsarkivet i Trondheim; Strathclyde Regional Archives in London; Tromsø Museum; United States Coast Guard stations at New York and Seattle; the Archives of the United States Military Academy at West Point; Vancouver (B.C.) Maritime Museum; and the Watt Library at Greenock.

Special efforts in our behalf were made (in Europe) by Ida Bull (Trondheim), Arne G. Edvardsen (Tromsø), Kari Helgesen (Trondheim), Björk Ingimundardóttir (Reykjavik), Tor L. Larsen (Tromsø), Sigurd Rødsten (Oslo), Caecilie Stang (Hamar), Else Marie Thorstvedt (Oslo), Ørnulv Vorren (Tromsø), and J. M. Wraight (London); and (in Canada and the United States) by Ashby Ahwinoan, Robert E. Baldwin, Rolf Baldwin, David Bradley, Adrian Bull, Claire C. Gardner, Elizabeth Gould, Ellen Jordal, Jacob Jordal, Joyce Justice, Leonard G. McCann, Michael E. Pilgrim, Della Price, R. L. Rausch, Amy J. Roberts, Matthew Robus, Teri N. Robus, Siegelinde Smith, and Lynn Wright. Among the many others who assisted were Deirdre Allan, the late M. A. Baldwin, D. T. Barriskill, DeAnne Blanton, Jane Brown, B. M. Derrick, Christin Gleeson, Milton O. Gustafson, A. M. Jackson, Gail L. Kerr, Jo Koch, John Koza, Ann Martin, Susan Miller, I. J. Monteith, Anne Møretrø, Stephanie Muntone, V. Perkins, Gregory J. Plunges, Kenneth J. Ross, the late E. E. Sacressen, G. Soumie, S. L. Srofe, Roberta Thomas, John K. Vandereedt, C. S. Viehweg, and C. Wildschut. James P. Nelson, who, with elegant precision translated Carl Johan's journal from Norwegian into English, supplied helpful literature.

We express our sincere thanks.

Carl Johan Sakariassen. 1960.

# Prologue

## From Dawson on the Klondike River....

during late summer and autumn of 1897 along with fresh tidings of the fortunes in gold being taken in the Yukon District came ominous messages that local supplies of food were critically inadequate for the winter months to come. Apprehension arose lest many of the miners, most of whom were citizens of the United States of America, would suffer privation, or would perish. Congress took quick action. The Department of War was directed to carry out with no delay a comprehensive plan of aid. Sledges loaded with provisions would be pulled overland to Dawson in February, by reindeer from Finnmark in Norway, and the deer would be managed on the trail by men also from that district.

To Congress and the secretary of war, Sheldon Jackson had proposed the approved plan of relief, and he was appointed to lead a mission to England and Norway to find means to fulfill the congressional mandate. William A. Kjellmann was sent to search the markets in Norway for reindeer and forage, to purchase appropriate animals and supplies, and to enlist men and women to accompany the deer to the Yukon. Daniel B. Devore was assigned to special duty as disburser of funds. In Europe, the three men together guided the course of the Yukon Relief Expedition, or the Reindeer Expedition, as it later was called.

On 4 February 1898, significantly behind schedule, the full company sailed for America from Bossekop, the major port of Finnmark. A few hours before the ship departed, a young Norwegian, Carl Johan Sakariassen, joined the mission. He began at once a personal journal, recording daily his observations as the Yukon Relief Expedition made its way across sea and continent, to Alaska and the Yukon.

While all components were being assembled and as the ship traveled westward, conditions in Dawson were changing. The miners were moving out, dispersing to settlements where provisions were not drastically short, and accordingly, in March 1898, the secretary of war declared that the crisis in the Yukon District had passed. Although the relief effort was cancelled, Sheldon Jackson's group continued on to Alaska as the Reindeer Expedition. Here, with Carl Johan's journal, is its story.

# Introduction

## I. Setting and Background

On Friday, 29 July 1898, the Bering Sea was almost calm after a week of strong northerly winds along the coast, and Captain Allen on the *Del Norte* could take his ship northeastward from St. Michael on a relatively easy trip to Unalakleet, in Alaska, a distance of about sixty miles. The two-masted schooner was carrying important cargo that included the annual stores and supplies to be off-loaded at the villages of Unalakleet and Golovin, and numerous passengers, some of whom were going to coastal sites to prospect for gold, even though the main excitement about mining was not in western Alaska in that month of July but was far to the east, in the Yukon Territory and adjacent areas in the District of Alaska. Among the passengers on the *Del Norte* was Sheldon Jackson, D.D., LL.D., who came aboard at St. Michael. For thirteen years Dr. Jackson had held the office of United States general agent of education in Alaska (a position in the Bureau of Education, U.S. Department of the Interior). On the *Del Norte* for the trip to Unalakleet, he was joining his special group—forty-two men and women (with six children) who were to disembark at Unalakleet and begin work on a major enterprise, constructing a new center in the lower Yukon River region to house and maintain people engaged in the reindeer industry and proposed for use also as a post in the overland federal mail run of western interior Alaska.[1] Most were employees of the U.S. Department of War as members of its mission—the Yukon Relief Expedition—to assist the miners in the Yukon Territory, but when that mission was cancelled, all personnel were assigned to work on the construction project—a part of the program with domesticated reindeer that Dr. Jackson, supported by the Bureau of Education, had begun some eight years earlier in an effort to establish vocational training schools and agricultural experiment stations (as existed elsewhere in the United States) for the benefit of the inhabitants of the District of Alaska.[2] Using the meager funds that the U.S. Congress had granted for education in the district, Jackson had managed to import reindeer from Siberia and incorporate training for their management within the curricula of mission schools in western Alaska. He had received vocal support from the highest officials of the federal administration in Washington, D.C., from senators and representatives in Congress, from his colleagues and other members of the Presbyterian Church of the United States in which he was held in high esteem, and from a segment of the general public. His friends in Alaska were fewer in number, but in July 1898 his critics were not widely active. For the latest project on the Unalakleet River, all of the essential tools and equipment, the manpower, and the stores and food stocks for months of sustenance for about eighty people were aboard the *Del Norte* and were to be carried in small boats onto the beach, when at last the ship would arrive.[3]

The arrival was on 31 July, and although it was Sunday, Dr. Jackson wanted to take advantage of the fine day. The surf was only a ripple along the shore, and the weather was perfect for unloading the equipment and supplies for the project as well as goods for the local people, in particular those to be delivered to the Swedish Covenant Mission Church at Unalakleet.[4] The *Del Norte* anchored offshore, about a mile and a half south of the little village, where a herd of reindeer had been resident for some months under the care of several families already employed in the reindeer husbandry program. Aboard the *Del Norte*, Jackson ordered all of his men to begin work at once, loading the tons of goods onto small boats, rowing to shore, and once there, unloading and hauling all to safety on land, away from the surf, which while gentle today might be tumultuous tomorrow. The reindeer herders and people with boats from the village came to help, including the men from Jackson's group who had arrived in Unalakleet from Seattle on the schooner *Louise J. Kenney* two days earlier; Jackson's trusted employee and companion William A. Kjellmann was among them.[5]

In April 1897, Kjellmann himself had selected the area a short distance from Unalakleet, upstream on the Unalakleet River, as ideal for reindeer; the countryside supported rich, dense lichen growth (reindeer moss) and extensive mixed forests of white spruce, birch, and aspen quite near the Bering Sea coast, and as a site for reindeer husbandry, it was relatively accessible by water transportation. Although the large ships had to anchor offshore and connect to the village via boats of the local residents, Unalakleet had an excellent, protected harbor for small vessels.

Just adjacent to the natural moorage cove was the mouth of the Unalakleet River which flowed into Norton Sound (eastern Bering Sea) from the low mountains on the northeast where lichens grew so abundantly that the slopes seemed perpetually white—in wintertime with snow, in summertime white with the glow of reindeer moss. The stream was shallow, but well navigable for river boats. It seemed that Kjellmann could scarcely have chosen a more fitting site in the area, nor one of more quiet, wild splendor.[6]

By midafternoon of 1 August, still with calm sea and minimal surf, all the supplies were on the beach, and the *Del Norte* headed off for Golovnin Bay. Aboard her was Dr. Jackson. Before he departed, he had appointed William Kjellmann to be leader of the men and women of the construction company, and to be general manager of the project, to plan, design, and supervise framing out of timber cut from the forests along the Unalakleet River the buildings for headquarters and housing for personnel of a new center for the reindeer industry, to be approximately eight miles upstream from Unalakleet village. Kjellmann began at once to organize the men (consulting with each and gaining acquiescence of each, to avoid complaints later on) into functional subgroups; in a general meeting, they discussed regulations, agreed upon issuance of food supplies, chose leaders for specific work details, and made ready the locally rented skin boats that would carry them and their supplies and equipment up the river as soon as weather permitted. Not until 5 August was the sea sufficiently calm again and the surf down. On that day the company set off, launching three boats holding fifteen men and laden with only a portion of the supplies. Entering the wide river mouth and using sails with a good wind from the southwest to help, the three boats made their way upstream.[7]

The boat in which Kjellmann was sitting was light and fast. He and his crew would arrive first at the site, a timing probably hoped for, inasmuch as he as general manager had planned a welcome ceremony for his men. Before the other boats drew ashore, Kjellmann walked up the slope, away from the river, to a tall spruce tree, chopped off some of the branches, and raised to its apex the Stars and Stripes flag of the United States of America. When every man had come ashore and all the boats were unloaded, Kjellmann assembled everyone up the slope beneath the flag, and shared toasts of schnapps with all. They cheered for the safe arrival of the company of men and of the equipment, and for the new center that they were now to build.

The builders and associated workers present that day on the wild northwestern bank of the Unalakleet River, and all who would join the company in the summer of 1898, were from northern Scandinavia—Finnmark Province of Norway. Carl Johan Sakariassen was one of that company. In 1897, in Norway, out of the local group of young men who were to serve in the national army, he had been called to the Royal Military Service. When he expressed an interest in joining the Yukon Relief Expedition, he was permitted to do so, but perhaps he received that authorization barely with time to travel to Alta to sign the necessary documents as a member of Sheldon Jackson's mission before he boarded the ship (the *Manitoban*) for North America at Bossekop (the port station of Alta) on 3 February 1898, along with 112 of his compatriots.

Carl Johan was then twenty-three years old, youngest son of Sakarias Johannessen and Brita Hendriksdatter of Fladstrand, the family home at the southern end of Langfjord (approximately 70° 2' N, 22° 15' E). He had attended the parochial (Lutheran) schools of Norway and became especially skilled in mathematics. He was strong and healthy in 1898, about six feet in height, smoked a pipe, and had a great love of good food and a pleasurable table. With deep attachment he cared for the family's homestead and the lands about it—the open slopes where he walked in summertime and skiied in winter; perhaps in truth he did not want to leave those places, nor his country. He had been a fisherman, sailing aboard vessels seeking cod and herring in the fjords and waters north of Hammerfest, beginning when he was about fifteen years old, and he felt as much at ease on the fjords as in the mountains, but when on 3 February he walked aboard the *Manitoban*, stowed his gear in quarters below deck, and looked about the old vessel with her cosmetic fresh paint, he felt a bit of apprehension. That evening he began a journal; for more than a year, almost every day was

Fladstrand, Norway. Leif Boland, grand-nephew of Carl Johan, could not be certain that the old-style structure was that of Carl Johan's parents, but he stated, "I think it is precisely like the house where he [Carl Johan] was born and grew up." The house and all the other buildings in the area were burned in autumn 1944 by soldiers of the German army, which occupied (1940–1945) almost the whole of Norway during the Second World War.

*—Photograph provided by Leif Boland.*

observed with written notation of events and personal impressions. He was not always content; sometimes he thought that the behavior of his companions was not appropriate, and sometimes again, that the leadership of his chief, Sheldon Jackson, was arbitrary and unfair.

Carl Johan was bilingual (Norwegian and Finnish), and had competent knowledge of Danish, Swedish and German—languages which during the time of the expedition were not really helpful to him. He later became fluent in English, but in 1898–1899, his lack of knowledge of that language was a definite disadvantage, perhaps contributing to less than complete understanding of some events. Nonetheless, that he strove for accuracy in his journal is apparent, and his observations were recorded plainly and in a straightforward manner, the writing no doubt also influenced by the often difficult conditions of travel and work. He noted in his papers that if his journal were ever to be published, he would hope for editorial attention. Accordingly, we modified some entries, mainly in word order and syntax. A few entries were omitted as not pertinent to the narrative of the expedition (for instance, a section, erroneous, on Alaska history, perhaps written without references during the winter of 1898–1899 or the year following).

But for those alterations, the journal is complete, and his day-to-day account is given herein.

On 6 August 1898, the day after landing at the site on the Unalakleet River, Carl Johan wrote:

> We contemplated this spot where we had arrived; what good it could offer us; why we were brought from Norway; and what Kjellmann had found up here, among the desolate, uninhabited hills by the banks of the river. That he should go just here, about eight miles from the sea, into the woods, to construct the main station for raising reindeer in Alaska, is something for which I still could not find a reason. The only thing I can think of is that enough kindling and timber are present in order to have houses for the winter—which could perhaps be a good enough reason to construct the station here. And with regard to food for the reindeer, there is moss everywhere one looked; but there could be many other reasons that I do not understand.

Probably, few people understood the complexities of the political and other factors that influenced the course of the Yukon Relief Expedition from its inception. It came to be but a small part of all that was being done to establish in Alaska a reindeer industry, about which certainty of purpose had prevailed for some years.

## II. The Cruise in Northern Alaska, 1890

The Department of the Interior's new center for reindeer husbandry near Unalakleet would be called Eaton Station, in honor of Dr. John Eaton, the United States commissioner of education, who during his term in that office (1870–1886) had sought from Congress the funds to establish schools and provide teachers in a general program of education for the children of the Alaskan people. In 1884, under Section 13 of the Organic Act, which gave structure for civil government in Alaska, the educational program in the District of Alaska was placed under the supervision of the secretary of the interior who was to administer an annual appropriation of $25,000 for that purpose; the next year, responsibility

Sheldon Jackson.

*—From an engraving by A. H. Ritchie, N. A.: in Sheldon Jackson's book,* Alaska, and Missions on the North Pacific Coast, *Dodd, Mead and Company, Publishers, New York, 1880.*

for education in the district was assigned to the commissioner of education. With assurance of federal funding, the commissioner established the position of general agent of education in Alaska, who would inspect, report, recommend, and carry out educational projects in the territory, and appointed Dr. Sheldon Jackson to hold that office. Dr. Jackson, as superintendent of Presbyterian Missions in the Territories, had been working since 1877 with and for the schools that his church had founded in southeastern Alaska, using funds donated by congregations of his own and other religious groups, nationwide, for the education of the Indians and Eskimos of the territory acquired by the United States from Russia some ten years earlier. In that private capacity, Jackson had been sending reports to the commissioner of education in Washington, D.C., for at least seven years. He had been quite successful in his requests to the various groups for financial aid for the schools during those years.

Prior to 1885, no federal funds were appropriated or available in any way for educational purposes in the district (often called a territory). But schools did exist there; the imperial government of

Russia continued to support seventeen, and the Alaska Commercial Company maintained one on St. Paul Island, one on St. George Island, one at Unalaska, and one (temporary) at Kodiak.[8] A few were funded by religious groups of the United States, and the Anglican Church of Canada had established a school in the upper Yukon River region before the year 1867. That the Presbyterian Church was operating schools at six villages and towns by 1885 was partly a result of Jackson's fund raising and recruitment of teachers. At the time, the area of greatest activity for the church groups was the southeastern coast of Alaska, which in geographic location was closest to the United States and British Columbia; Jackson was well acquainted with the settlements of the coast, as he had often visited and worked there in summer and early fall. After federal funds became available and Jackson as general agent of education was able to expand educational programs, attention spread to consider all areas, including those of the Arctic Coast. Obedient to the mandate of his office, he began as soon as he could to make educational surveys of the existing schools and to determine placement of additional ones for residents in the territory north of the southeastern Panhandle.[9]

His first inspections of the settlements and assessments of educational requirements along the coasts of the Bering Sea and the Arctic Ocean became possible when he was granted permission by the secretary of the treasury to travel as an official passenger on the U.S. Revenue Cutter *Bear*, Captain Michael A. Healy commanding, on her annual cruise in northern waters in 1890, on which cruise the cutter would stop at the major villages in Alaska and would visit the Siberian Coast as well. A cordial association between Captain Healy and Dr. Jackson began during the summer's travels. Each had a federal purpose and a mission to aid citizens of the United States and to consider positively the indigenous people—who were not citizens but Jackson and Healy were certain of their need for education and "civilizing," which included urging to Christianity (as expressed by Jackson, at least). With an element of patronage, they were concerned about the welfare of the Alaskans.[10]

Jackson had respect and admiration for Captain Healy.[11] The captain of the *Bear* had been sailing the northern seas since 1868. At a time when shipwrecks were commonplace and almost expected, given the risks and dangers involved for ships in the uncharted coastal waters and the moving ice packs of the northern seas, Captain Healy had never lost a vessel. Every year if ice and weather conditions permitted the *Bear* would call at all settlements on the islands and coastal mainland, as a representative of the single arm of the United States government present and responsible for order in waters along the coast, carrying the mail to and from the far north, and charged with the duty to deal with any activity, domestic or foreign, that was contrary to United States laws, such as smuggling and nonpayment of customs duties, selling or distributing liquor, killing fur seals on the open seas, and crimes of passion, murder, or theft, for any of which the individual Alaskan could be detained or arrested by Captain Healy's men and locked up aboard the *Bear*, to be appropriately released only at the first port where competent authority resided for trial and judgment. Revenue cutters were vessels fitted for life saving as well, searching for and rescuing disabled vessels, stranded passengers and shipwrecked sailors, and assisting others in need along the coast.

Many of the inhabitants of the northern villages seemed to Jackson to be in difficulty, approaching starvation due to the scarcity of their usual food resources (seals, walrus, and whales), and otherwise in a parlous state of health and general welfare. Captain Healy was well aware of those difficulties, but the U.S. Revenue Cutter Service was not obligated to perform charity, and in any case, no substantial improvements would have been forthcoming out of dispensing the simple and limited stores of the cutters. The summer's experience on the *Bear* and in many northern communities was like no other for Dr. Jackson. He visited nineteen settlements along the Alaska coastline in Bering Sea and the Arctic Ocean and four settlements in Siberia, and collected information concerning the best sites for the establishment of schools in Alaska. On numerous occasions he was invited to join Captain Healy and Mrs. Healy in the captain's quarters to have a meal or to talk; they discussed the grievous conditions of the Yup'ik and Iñupiaq populations. Jackson concluded (as Captain Healy had already judged) that the activities of commercial hunters from areas south of Alaska had directly affected the aboriginal peoples by depleting their food supply.[12]

Whaling vessels had been working north of Bering Strait since about 1848, each year taking whales in greater number. Two-thirds of the total catch of bowhead whales, to the year 1914, had been killed by about 1870. The populations of that species became less accessible, and by 1890, voyages far along the pack ice east of Point Barrow were undertaken in order to find those animals. Walrus also were eagerly sought by the whaling fleets, so that by about 1880, they were scarce.[13] Numbers of other marine mammals were observed to be far below those existing earlier.[14] That the whalers provided (illegally) the local people with alcoholic spirits seemed to interact disastrously with effects of the depredations on arctic sea mammals; not only was the people's food supply depleted by the commercial whalers, but the capabilities of the indigenous hunters were adversely affected through use of the alcohol. The combination of factors was thought to contribute significantly to bring about the poverty and misery Jackson saw at many settlements along the Arctic Coast, or thought he saw, for there are occasional contradictions to that judgment in his own reports. During summer of 1890, as ordered, the crew of the Revenue Cutter *Bear* dumped many a barrel of liquor into the sea and destroyed other stores of alcoholic spirits that they could find on the whaling vessels and ashore. The whaling industry itself was not to be controlled, nor was such a radical course favored. Few raised a protest against the legitimate business of the whaling fleets, which was providing the highly desired and profitable whalebone (baleen) and oil.

As a result of his observations during the summer's cruise, including the visit in Siberia with Chukchi reindeer herders, and of his talks with Captain Healy, Dr. Jackson became convinced that the United States government should begin to import domesticated reindeer from Eurasia in order to establish reindeer husbandry amongst the Yup'ik and Iñupiaq peoples. He believed that with reindeer herds, as were maintained in Siberia, no longer would those North American peoples be dependent for their food on the ever-scarcer marine mammals. The reindeer would also serve as a means of transport and travel with sleighs. He believed as well that caring for the deer and becoming herders would tend to be civilizing, probably Christianizing, influences for the so-called untaught peoples of the Alaska north. Jackson considered that establishment of a program of reindeer husbandry would be a justifiable and appropriate extension to Alaska of the force of the Agricultural Acts of 1862, 1887, and 1890, pertaining to agricultural experiment stations, vocational training, and means of federal support.[15]

After the summer's cruise, Jackson communicated the proposal to the commissioner of education (then Dr. William T. Harris) and others in the federal departments, to members of Congress, and to colleagues in the Presbyterian Church; many judged his plan to be sound, faultless, even ideal, and attractive in that it was economically feasible, for the nucleus of a reindeer herd could be established in Alaska very cheaply with reindeer purchased in Siberia, probably at a cost of less than seven dollars per animal.[9] But Congress was not ready to give the Alaskans the benefit of the agricultural acts. Eventually, they allocated federal funds to assist the husbandry program through an annual appropriation to the

Bureau of Education.[16] Encouraged by the verbal if not financial support amongst officials of the government in Washington, D.C., Jackson appealed to the public and accumulated sufficient money to buy sixteen deer in 1891; Secretary of the Treasury William Windon instructed Captain Healy to assist in locating the Siberian herds and transporting the deer from the Asian coast to Alaska. The Russian government assured Secretary of State James G. Blaine that local officials in Siberia would help as well. During the years 1892 through 1895, 538 domesticated reindeer were introduced; the number increased through reproduction to 1466 by 1897, and herds had been established at Golovin, Cape Nome, Cape Prince of Wales, Teller Station, and Fort Adams (near the junction of the Tanana River with the Yukon River in interior Alaska). The first congressional appropriation, six thousand dollars, to apply to the program was approved in 1894.

## III. The Summer and Autumn of 1897

In late winter of 1897, Commissioner of Education Harris asked Sheldon Jackson to make a reconnaissance during the coming summer along the Yukon River, from St. Michael near the river's mouth to Dawson City, Yukon District. St. Michael was approximately 125 miles north of the main outlet of the river, and as the major port in the region was the western depot for river traffic up and down the Yukon. The river was growing in importance as a water route to the recently found gold fields at Dawson City and elsewhere in that area, and Dr. Harris wanted to have information about the prospects for use of reindeer in interior Alaska as a means of transportation, not to replace the river but to add further mobility, especially during winter. Secretary of Agriculture James Wilson also was interested in the Interior as an area where farming might be introduced, and he requested Dr. Jackson to evaluate conditions with that in mind. Jackson, accompanied by William Kjellmann, left St. Michael on 3 July 1897 aboard the river steamer *Portus B. Weare*. That summer, the water level of the river was exceedingly low, and few of the larger steamers were able to navigate from the sea with full loads as far as Dawson City and the other mining camps.[17] At Andreafski Village, about 215 miles upstream from St. Michael, provisions sent in spring and early summer from San Francisco and Seattle for the miners' camps were piling up; even greater was the mass of equipment and food at the port of St. Michael, all of which was to have been in Dawson City by September before freeze-up on the river. To Jackson and Kjellmann, that most of the needed supplies would never reach the gold camps in 1897 became obvious as they traveled in the Yukon's dwindling channels, with sandbars exposed to either side of the *Portus B. Weare*; the little river steamer made it to Dawson, but on the return trip, she ran into a bar just below surface of the water and was grounded fast, far from any settlement. While they waited for another vessel to happen along, some of the passengers and crew spent time fishing or used the ship's boat to go ashore for walks; Jackson made observations and collections of vegetation in the taiga along the river. After nineteen days, the steamer *J. J. Healy*, en route downstream, stopped to assist, and was able to take on most of the stranded passengers.[18] They arrived back at St. Michael on 24 August, and Jackson immediately reported his observations on the condition of the river to the Department of the Interior. Already by the end of August the news from the Yukon District carried predictions of food shortages during the coming winter for the population at Dawson City and vicinity; some thousands of people had gathered there, to look for gold or to enrich themselves in other ways. In September, President McKinley and members of Congress were receiving petitions and letters, asking for federal intervention to avoid disaster for the miners. With Dr. Jackson's advice the president approved of the plan developed, to drive reindeer from the government's herds in western Alaska overland to Dawson City, pulling sledges full of supplies for the camps. In orders sent on 22 September 1897 by telegraph, the secretary of the interior assigned Kjellmann the job of coordinating the program in the field, selecting reindeer and building sledges, and the project was to be under the direction of officers of the Department of War (U.S. Army), stationed at St. Michael.[19]

Kjellmann probably did not receive the telegram until the next month, since it was carried to St. Michael by ship. At once in October he began the organizational work, but new orders from the Bureau of Education in Washington, D.C., in effect cancelled the plan (which in actuality had been laid prematurely since shortages had not yet been confirmed in September). He himself was reassigned, called to Washington, and there stood by while the administration considered what was to be done about a second problem that had arisen in Alaska. In the Arctic Ocean near Point Barrow, a number of ships (no one was certain, at first, how many) of the whaling fleet were reported to be caught fast in ice, with provisions much too short to last throughout the long arctic winter.

Like the miners in Yukon District, the ships' crews were facing privation, at best, or perhaps starvation. Not until November was doubt fully dispelled concerning the respective shortages.

The first definite indication of potential trouble for the whalers came at the end of October. Reverend Vene C. Gambell and Mrs. Nellie F. Gambell, on leave of absence from the Presbyterian mission school on St. Lawrence Island, were aboard the steamer *Bristol* when she arrived in Nanaimo, British Columbia, on 27 October 1897. The Gambells brought fresh news from the Arctic: they had learned that five ships were in the ice, offshore from Point Barrow.[20] The information was not especially alarming; whaling ships frequently overwintered in the Arctic without damage or hardship for the men (and occasionally women) aboard, although usually such a time was spent in a coastal lagoon or a sheltered area farther to the east, such as the waters near Herschel Island. But during the first week of November, another report came from the crew of the steam brigantine *Karluk*, on arrival at St. Michael: offshore from Barrow, eight whaling vessels were lodged in the pack ice, some severely damaged, and food was far less than would be needed for survival of the shipwrecked men.[21] The news was carried on the front pages of the major newspapers west and east, the *San Francisco Chronicle* and the *New York Times*, and accounts appeared almost daily in other journals as well, as information accumulated. Evidently with as much concern as that expressed for the miners at Dawson, people appealed to President McKinley and the Congress, asking that the government act to save the whalemen from starvation.[22]

Eight years after the cruise of the *Bear* in the Bering Sea and Arctic Ocean, on which Sheldon Jackson was an official passenger (1890), the ever-present, effective arm of the federal government in Alaska remained the Department of the Treasury and its revenue cutters, and President McKinley, following a conference with his cabinet and others, turned to that department for help in the crisis of the arctic whaling fleet.[23] Again men of experience in the Arctic, of whom Dr. Jackson was one, were consulted. Secretary of the Treasury Lyman J. Gage ordered the *Bear* (Captain Francis Tuttle commanding) to make ready with provisions and a plan to deliver them to Barrow, and to be underway as soon as possible. The delivery of supplies was to be in two parts: immediately (in late November 1897), the *Bear* was to proceed as far north along the Alaska coast as ice permitted in order to land a party to go overland with dogs and sledges to reach the reindeer herds of the government, the religious/educational missions, and individual owners, from which herds a number of deer would be chosen and driven northward overland to Barrow. Then, when navigation through the sea ice became feasible (seven months hence—in July or August 1898), the *Bear* was to make its way to Barrow and find and carry south any surviving whaling crewmen. The cutter sailed from Seattle on 27 November 1897 and traveled in the Bering Sea as far as St. Lawrence Island, but was forced by conditions of the ice to turn back; the overland party of three men of the Revenue Cutter Service (1st Lieutenant David H. Jarvis, 2nd Lieutenant Ellsworth P. Bertholf, and Surgeon Samuel J. Call) and their supplies were landed on the coast near the village of Tanunak. The plan as formulated was carried out successfully, with assistance along the way from many people of several communities, and on the drive to Barrow with participation of several Iñupiaq men under leadership of William T. Lopp of the mission school of the Congregational Church at Cape Prince of Wales, and Charlie Antisarlook, owner of a large herd at the Cape. After an overland journey of approximately 1500 miles (for Jarvis, Bertholf, and Call, and about half that distance for the others), the men arrived at Barrow with 382 reindeer on 30 March 1898; the *Bear* arrived on 28 July 1898, to take aboard crews of four wrecked vessels.[24]

But in December 1897, the American people, the president, and Congress could only hope that the men of the whaling fleet would be able to survive until such help arrived in late winter or early spring; for half a year, no one in the United States would receive reliable information about the crews or the attempt to rescue them. In the meantime, attention had to be directed to the people at the gold fields of the upper Yukon River valley. To assist them, far fewer reindeer were now available in western Alaska, and to draw many more from those herds beyond the number sent overland to Barrow would be difficult and conceivably would imperil the stocks that the Bureau of Education had managed to build for use of the local people. If additional reindeer were to be required as a means of transport of supplies to the Yukon, or for food there, a new source of the animals would have to be found.

# IV. The Congressional Debates

Along with news about shipments of gold (the last for the year) received at St. Michael and en route to Seattle and San Francisco came reports in September 1897 from the upper Yukon River valley near Dawson City that distress due to lack of food prevailed in the mining camps, and the subarctic winter was approaching. There, where the Klondike River joins the Yukon (then and afterwards often called simply "the Klondike"), of the eight thousand people present, some five thousand were reported to be without adequate provisions, in part because the unusually low water level in the Yukon had prevented river steamers with stores from entering the upper valley.[25] Willing as they might have been to spend their gold for supplies, the miners could find little to buy at any price, or so it was said. The reports of distress were conflicting; indeed, the number of inhabitants in and around Dawson City was known only as a close estimate since no one had made a count. Nonetheless, in Oregon, Washington, and California especially, many were convinced that the lives of thousands of people were at risk in the Klondike, and in petitions to President McKinley and the Congress they asked that the federal government send aid to the miners. The president and his cabinet at once attempted to determine the extent of need; Secretary of War R. A. Alger already (in August) had directed that Captain P. H. Ray and Lieutenant W. P. Richardson of the Eighth United States Infantry stationed at St. Michael were to go to the upper Yukon to investigate and report on the conditions of the several communities and their needs, military as well as civil, including food stocks. By the end of November, Alger had received from Captain Ray only three communications, sent from Circle City and Fort Yukon. He had hoped for more, but based on those dispatches and on testimonials of recent travelers from Dawson City, the secretary of war advised the House and Senate in early December that unless supplies were sent to the Klondike, many would suffer extreme want.[26]

Obviously, succor for the miners was going to cost a great deal more money than the original arrangement based on use of local reindeer already belonging to the federal government. That plan (which in September William Kjellmann and Lieutenant Colonel G. M. Randall of the Eighth Infantry at St. Michael had been ordered to implement) had been cancelled. Unfortunately, the number of government-owned animals was too small to meet the demands presented by two concurrent emergencies—the whalers on the Arctic Coast and the miners in the Klondike, alike perhaps facing starvation. A new plan of action was needed, to which it was hoped Congress would give generous support.[27]

In the Senate on 9 December 1897, George W. McBride (senator from Oregon) introduced a joint resolution, S.R. 72, appropriating $250,000 for assistance to the people of the upper valley of the Yukon and authorizing the secretary of war to administer the fund and select the means of transport of supplies to which end U.S. military forces and private individuals could be employed. He asked that immediate consideration of the resolution be given (implying his hope for passage by the Senate's acting as a committee of the whole), and when Senator Francis M. Cockrell (Missouri) objected on the basis of lack of official information about actual conditions in the Yukon area, Senator McBride presented a petition sent to President McKinley by the chamber of commerce of Portland, Oregon, as well as remarks of credible men who had arrived in Seattle from Dawson City two weeks earlier.[28] To Senator William E. Chandler (New Hampshire), that information was convincing and sufficient for action by the Senate, but Senator Cockrell's request governed the procedure on the floor; to substantiate further the critical need for aid, the Senate would ask the secretary of war for a report about conditions in the mining districts of the Yukon.

Sheldon Jackson was then in Washington, D.C. Undoubtedly, Secretary of War Russell A. Alger called him for consultation, and the Senate received Alger's report by 15 December.[29] Two days later, Senator McBride presented his colleagues an amended S.R. 72, the language of which directed that reindeer be purchased abroad and reindeer drivers be employed, all to be transported to the United States under the authority of the secretary of war. To the motion for referral of the bill to the Committee on Military Affairs, Senator Thomas H. Carter (Montana) courteously asked acquiescence of Senator McBride. McBride made no formal objection, but cited recent precedents wherein resolutions for relief had been passed without referral to committee. His firm certainty about the need for assistance in the Yukon, without delay, probably was a positive factor in the speed with which the committee's favorable report on the resolution was accepted. S.R. 72 was approved by the Senate on 16 December 1897.

In the meantime, Representative Joseph G. Cannon (Illinois) had prepared a similar bill (H.R. 5173) for the House, and while his measure would allocate less ($175,000), its language was more specific, as Cannon had included detail provided (upon request) by Sheldon Jackson. The clerk of the House was asked to read Jackson's memorandum (dated 15 December) of five paragraphs, masterfully composed, plus addenda about geography of Alaska and estimates of costs, all of which gave background relative to

H.R. 5173. Congressman Cannon addressed the House: "Mr. Speaker, gentlemen know who Dr. Sheldon Jackson is. This proposition received the approval also of the War Department...."[30] Representative Joseph W. Bailey (Texas) was the first to object to Cannon's request for immediate unanimous consent of the bill, as he could not understand how the miners could be called poverty stricken amidst the marvelous richness reported of the Klondike, and he suggested that a most undesirable result of any federal aid would be that the U.S. government was going to become a storekeeper, selling and bartering provisions, when it ought in this instance to give outright. He declared that he did not intend to support the measure, in any case, but he thought that charity ought to be voluntary, and if not, it was a matter for state, not federal, government, and he gave the example of his own state of Texas, whose legislature was generous and gave without qualification to the needy, and thus the county poorhouses of Texas were not full, and he assumed that every state legislature would do the same. His objection provided an appropriate occasion to remind the House,

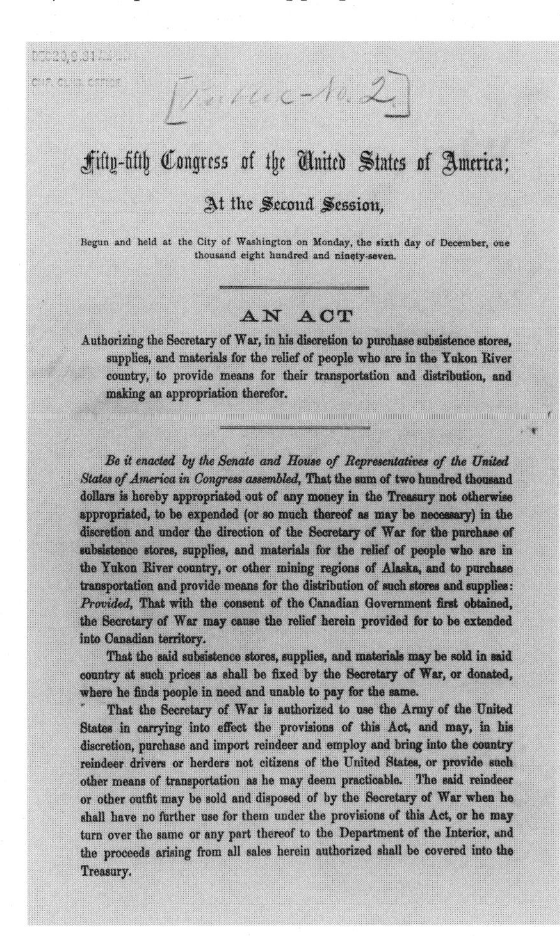

Fifty-fifth Congress of the United States of America;

At the Second Session,

Begun and held at the City of Washington on Monday, the sixth day of December, one thousand eight hundred and ninety-seven.

AN ACT

Authorizing the Secretary of War, in his discretion to purchase subsistence stores, supplies, and materials for the relief of people who are in the Yukon River country, to provide means for their transportation and distribution, and making an appropriation therefor.

*Be it enacted by the Senate and House of Representatives of the United States of America in Congress assembled,* That the sum of two hundred thousand dollars is hereby appropriated out of any money in the Treasury not otherwise appropriated, to be expended (or so much thereof as may be necessary) in the discretion and under the direction of the Secretary of War for the purchase of subsistence stores, supplies, and materials for the relief of people who are in the Yukon River country, or other mining regions of Alaska, and to purchase transportation and provide means for the distribution of such stores and supplies: *Provided,* That with the consent of the Canadian Government first obtained, the Secretary of War may cause the relief herein provided for to be extended into Canadian territory.

That the said subsistence stores, supplies, and materials may be sold in said country at such prices as shall be fixed by the Secretary of War, or donated, where he finds people in need and unable to pay for the same.

That the Secretary of War is authorized to use the Army of the United States in carrying into effect the provisions of this Act, and may, in his discretion, purchase and import reindeer and employ and bring into the country reindeer drivers or herders not citizens of the United States, or provide such other means of transportation as he may deem practicable. The said reindeer or other outfit may be sold and disposed of by the Secretary of War when he shall have no further use for them under the provisions of this Act, or he may turn over the same or any part thereof to the Department of the Interior, and the proceeds arising from all sales herein authorized shall be covered into the Treasury.

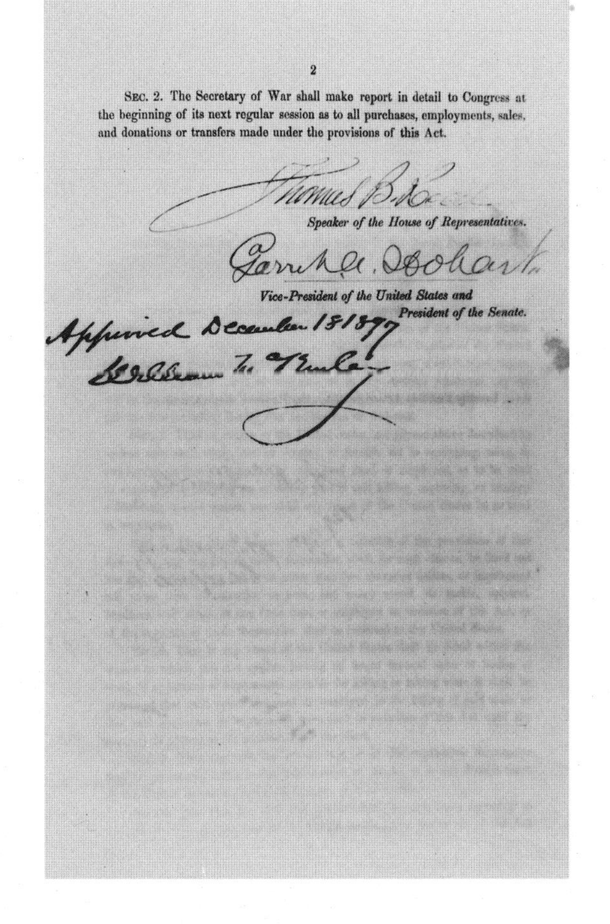

2

SEC. 2. The Secretary of War shall make report in detail to Congress at the beginning of its next regular session as to all purchases, employments, sales, and donations or transfers made under the provisions of this Act.

*Speaker of the House of Representatives.*

*Vice-President of the United States and President of the Senate.*

Approved December 18/1897

The act enabling the Yukon Relief Expedition: House Resolution 5173.

*—Photograph: National Archives, General Records of the U.S. Government, RG 11.*

which Representative William S. Knox (Massachusetts) immediately did, that the United States Congress was itself the government for the American people in the northern gold fields. Directing his remarks especially to Congressman Bailey, Representative Joseph D. Sayers (Texas) expanded the reminder, to point out that Alaska was a territory with neither state organization nor county organizations, and the federal government did not recognize any governing body there except itself; the people of Alaska only had improvised their own local arrangements.[31] Cogent questions were raised by other congressmen. Representative Jerry Simpson (Kansas) wondered how the miners could survive until 1 March 1898, which was the estimated date of departure of the proposed reindeer trains from Dyea, in southeastern Alaska, and how the deer would manage without moss, their necessary food, in the boggy, or ice- and snow-covered country over which they would be driven. Representative John F. Lacey (Iowa) wondered in what condition to haul loads of provisions the deer would be upon landing at Juneau, after a journey halfway round the world, for the animals were to be purchased in Lapland.[32] Representative Charles K. Wheeler (Kentucky) asked whether the majority of the miners were not in fact on British territory. To that inquiry, Mr. Cannon responded that indeed many of them were, but "...they are American citizens and if they are there starving, whether they are British or whether they are American, in the presence of starving and suffering humanity we can not fail to act in this matter," and the House of Representatives applauded.[33] Mr. Wheeler also was uneasy about the prospect that the United States would be establishing warehouses with goods to be sold, and acting as well in a country where Great Britain, not the United States, had sovereignty. Representative Sayers' response brought a second applause in the House: the knowledge that American citizens were suffering gave Congress the right to send aid to them wherever in the world, on territory of any government, they might be.[34]

Following Dr. Jackson's testimony, changes were made in the language of the bills in both houses. Delay was discouraged so that a conference bill could be passed before Congress adjourned for the Christmastime recess. One last point was that of Senator Augustus O. Bacon (Georgia), who wanted some restriction on record, to make known that the United States government was not going to have a general policy of care for the improvident, in case vast numbers of people went north in the future without appropriate supplies for themselves.

On 18 December, the Committee of Conference presented an amended H.R. 5173, allocating $200,000 for relief of the people in the Yukon River country. Vice President Garret A. Hobart in the Senate and Speaker Thomas B. Reed in the House signed the final version shortly before adjournment in the afternoon of that day. President McKinley gave his official approval on 7 January 1898.[35]

Notwithstanding the earnest discussion that took place in Congress concerning the proposal for federal aid, that either the Senate or the House would refuse to approve the measure seemed highly unlikely, for in Congress and throughout the United States, a perception prevailed (but was not specifically mentioned in the congressional debates) that the suffering miners of the Klondike were contributing positively and significantly to the economic wealth of the United States, as indeed it would seem that they were.[36] Even Representative Bailey of Texas, who complained at length about governmental charity, stated to his colleagues on the floor of the House that after all, the sum of money appropriated to help the destitute miners was not very great.[37]

Upon passage of H.R. 5173, the Department of War at once arranged for the travel of Dr. Jackson to Europe, to be accompanied by an officer authorized to disburse funds necessary to carry out the mandate. The accompanying officer was to be Lieutenant Daniel B. Devore, Twenty-third Infantry, United States Army.

## V. The Journey to London

As if coordinated with the date on which Congress authorized the Yukon Relief Expedition, William Kjellmann had arrived in Finnmark District, northern Norway, at Alta, a trading center of Norwegians, Saami, and Finns.[38] The records are obscure, but evidently his chief (Sheldon Jackson) had been convinced or assured that funds would be appropriated for purchase of reindeer there; already in late November Dr. Jackson had begun to seek information from suppliers (European commodities brokers, including those dealing in reindeer livestock in the Scandinavian countries) as to cost of purchase and shipment.[39] Such advance inquiry was prudent, or was suggested to him as appropriate, inasmuch as he was shortly to be called to advise Congress and the secretary of war concerning the problem of supplies for the Klondike miners. After congressional appropriation of funds and approval of the plan to import deer and permit entrance of reindeer managers into the United States, Jackson officially informed Secretary Alger that since Kjellmann was already in Norway, preliminary inquiries there about availability of suitable reindeer could begin at once, and Alger asked on 22 December by cablegram to Kjellmann how soon six hundred deer could be shipped and at what price, to which he received reply from Alta that five hundred could be sent in mid-February 1898. Alger probably understood that an estimate of cost was not given since Kjellmann had no authority to bargain with or name any amount to prospective sellers; the greatest concern of the secretary of war in any case was not of the cost but of the time of shipment, and he immediately cabled in response that the animals must be on their way to America by 20 January. As for money, evidently Kjellmann had only heard that some might be forthcoming from the federal government, and he had needed to borrow a thousand kroner in order to pay expenses in Finnmark.[40] If funds were assured, determination of the terms of purchase would be with concurrence on arrival in Norway of Dr. Jackson and the officer of disbursement of the War Department, Lieutenant Devore, whom Secretary Alger had specially assigned to be in charge of a portion of the $200,000 that Congress had given for relief of the gold miners.

Dr. Jackson received his own orders on 23 December, following precise and correct governmental procedure in which the secretary of war made official request of the Department of the Interior for the services of its general agent of education in Alaska; upon receipt, Secretary of Interior C. N. Bliss responded directly to Dr. Jackson:

> In compliance with the request of the Secretary of War, dated to-day, you are hereby directed to report to him for temporary duty in connection with the duties enjoined

Warrant No. 1913, Department of the Treasury (24 December 1897): The secretary of war assigned responsibility for these funds (one-third of the congressional grant) to Lieutenant Devore.

*—Photograph: National Archives, Accounting Offices of the Department of the Treasury, 18E-4, R-4, B396.*

WAR ACCOUNTABLE WARRANT.

UNITED STATES TREASURY DEPARTMENT

No 1913

$60000$

To the Treasurer of the United States. Washington, D.C. December 24 1897.

Pay to Treasurer U.S. for credit of Lieut. D. B. Devore, Special Disbg. agent for War Dept. or order

Sixty thousand Dollars $60,000.00

The Treasurer U.S. Washington will pay this warrant

Ellis H. Roberts, Treasurer.

Secretary.

No 7477

Comptroller.

P.O. Address Stockholm, Sweden, (Care U.S. Minister)

> by the act of Congress approved on the 18th instant, entitled 'An act authorizing the Secretary of War, in his discretion, to purchase subsistence stores, supplies, and materials for the relief of people who are in the Yukon River country, to provide means for their transportation and distribution, and making an appropriation therefor'.[41]

On that day also, the secretary of war sent specific directions to Dr. Jackson, who was now Special Agent, Department of War:

> You are hereby instructed to proceed at once, in company with Lieutenant D. B. Devore, United States Army, to Norway and Sweden, to purchase 500 reindeer broken to harness, together with full outfit for hauling or carrying supplies to the Yukon Valley, and also to provide such means as may be necessary for shipment to this country as soon as purchased, remembering that it is essential that they should be landed here at the furthest by the 15th of February. These instructions are also to cover the employment and bringing into this country of reindeer drivers.
>
> Lieutenant Devore of the Army will accompany you as Disbursing Agent and will provide the necessary funds for all expenses and purchases in connection with the procurement and shipment of the reindeer. I wish you and Lieutenant Devore to confer on all matters pertaining to the purchase of these reindeer and the shipping of them.
>
> These instructions are for Lieutenant Devore, as well as yourself.

Lieutenant Devore was military aide to Secretary of War Alger, and as such he received no special orders other than his assignment as given in the directive to Dr. Jackson.[42]

Provided with an official passport and a letter of recommendation from the Department of State (signed by Secretary of State John Sherman), Dr. Jackson walked onto the steamship *Lucania* on the afternoon of 24 December 1897 in New York harbor.[43] Lieutenant Devore arrived just in time before the ship departed for England. Not many passengers were aboard for that Christmastime sailing to Liverpool, and Dr. Jackson found that he had been given a stateroom all to himself. He was grateful, for he was exhausted from the stressful sessions at Washington. Letters from his family arrived for him, perhaps the only pleasant event for several days.

Beginning when the *Lucania* left New York early in the morning on Christmas Day and during the next two months of travel, Dr. Jackson kept records of many of his experiences and observations, and meticulous accounts of cash that he spent en route.[44] Two days at sea, he noted with relief that except for Lieutenant Devore, none of the other passengers, who mostly were British citizens returning home, knew who he was; no one asked for a lecture about missions in Alaska or gold in the Klondike. The anonymity of course could not last long, for he and Devore sat at the purser's table, and on 30 December he was speaking in the ship's hall, evidently to the great interest of the audience, among whom were Colonel Theodore A. Dodge and Mrs. Dodge, who later treated him and Lieutenant Devore to a fine dinner in London.[45] During the Atlantic crossing, Jackson found Devore to be a congenial companion, which must have been a second source of relief, in view of their assignment; together they would have to accomplish much in a relatively short time in strange lands. They hoped that their work would not fall everywhere into a pattern of delays such as they expected at their very first stop, England, where most people were still celebrating the holiday season. They would be in London on New Year's Day, but they intended to begin at once on that day the search for a ship to carry five hundred reindeer and a good quantity of reindeer moss to America.

News of the expedition had preceded the arrival of the *Lucania* in Liverpool on 31 December, and representatives of the White Star Line were waiting by the customs' section, eager to talk and sign a contract for use of one of their vessels to transport the deer but not willing that afternoon to name a price. Jackson and Devore listened, and took the Midlands train for London, where at the Cecil Hotel they at last found rooms of reasonable comfort and also a telegram from Kjellmann, requesting funds. It was by then midnight. Dr. Jackson began the New Year in prayer. He was up early on New Year's Day 1898, sat at light breakfast service at 7A.M., read the morning papers, and waited until Devore had finished breakfast—the main one served after 8 A.M. Then they set out by cab for Morgan Company's Bank to arrive before it closed at noon. Lieutenant Devore, with his letter of credit, succeeded in depositing money on account, with the intent to fulfill Kjellmann's urgent request by transmission by wire of funds to him in northern Norway, only to be told at the bank that in Europe, money was transferred from place to place as necessary by checks in the mails. With Kjellmann's telegraphed message in one hand saying that he was without cash and in the other, the order of the secretary of war for five hundred reindeer to be on board a ship within three weeks, Jackson and Devore did not question the bank's information but they did doubt that funds sent by mail

Daniel B. Devore: 1885.
*—Photograph by Pach Brothers, courtesy of the Archives of the United States Military Academy, Special Collections of the Library.*

would reach Kjellmann in time to enable him to buy or reserve the required animals to meet the schedule indicated by the War Department.[46]

Clearly, it seemed that either Jackson or Devore could more certainly deliver the cash personally to Kjellmann, as the time for the journey would be no more than that of a letter by mail. Since very shortly both of them were to travel to northern Norway, they agreed that morning that one would go in advance of the other; in two days, Jackson would leave for Finnmark while Devore would remain in England, and they would meet in Norway after Devore concluded negotiations for the steamship. They were placing themselves at risk of disregarding Secretary of War Alger's orders that they consult on all matters concerning their mission, but if that was perceived as a problem, Jackson made no mention of such a default. To him and evidently to Devore, the decision to proceed separately was well based and reasonable, and probably everything would have been accomplished without the slightest trouble but for having to treat with the local brokerage agency. The firm of Mocatta, Jansen and Company, Foreign Produce Brokers of London with connections in Trondheim, had been in contact via cablegrams with Secretary Alger, offering procurement of reindeer and moss and all equipment, and was avidly seeking a contract. When Jackson and Devore returned to the Cecil Hotel, Max Jansen, representing the firm, was waiting for them, and together they went to the Adelphi Restaurant for dinner and discussion. If Jansen paid the bill, the Americans were at once beholden, albeit only for a meal.

In the afternoon, Jackson and Devore called at the American Embassy, but Ambassador John Hay and his family were away for the holiday weekend.[47] Hay with his interest in affairs worldwide might have enjoyed meeting with his two compatriots, who could tell him much about Alaska, which conceivably would at least have entertained him, for he had been once rather closely associated with Secretary of State Seward in Lincoln's cabinet. He probably could have provided some good advice for the expedition, as well. The ambassador would have occasion to become acquainted with them, not through a sociable meeting as hoped for and as courtesy dictated to Jackson and Devore on this New Year's Day, but later on when as secretary of state, he had to deal with their eventual, unpleasant involvement with Max Jansen and Mocatta, Jansen and Company. Two days hence, Jackson and Devore again met Jansen, who evidently pressed so aggressively for a contract that Jackson felt ill at ease and suspected the broker's motives. At the time, they signed no binding documents with Jansen, although they did express permission that his company could assist them in some details of their mission. Jackson spent his last day in London attending church services.[48]

## VI. London to Norway

Early on Monday, 3 January 1898, Jackson and Devore called at the Atlantic Transport Steamship Company, hoping to settle at once the question of a ship suitable for reindeer as well as attendant passengers, but they received instead some discouraging news. An outbreak of contagious disease in cattle had occurred in Sweden, and the British Department of Agriculture had declared an embargo relating to any ship, domestic or foreign, that was carrying or had been exposed to livestock or raw products applying to livestock originating in any of the Scandinavian countries. To such vessels for a period of time then undeterminable, all British ports were closed, or a twenty-one-day quarantine was interposed before entry would be permitted.[49] For the Yukon Relief Mission, quarantine would be so costly in time and money that without question it would have to be avoided, although the route of the chartered ship, carrying reindeer from Norway, had been planned to include a brief stopover at a British port in order to ease for all aboard the long trans-Atlantic journey. The news was almost calamitous. With his plans completely upset, Jackson called again at the American Embassy. Ambassador Hay had not yet returned, but the man on duty, the embassy's chief clerk, suggested that the problem be discussed with an authority at the British Department of Agriculture. There, receiving Dr. Jackson with courtesy and much interest in the plight of the gold miners of the Yukon (which was, after all, in the British Dominion of Canada and the conditions at Dawson City might have been of as much concern to Britain as to the United States), the first assistant secretary of agriculture proposed that a vessel not lately employed in any way in livestock commerce be engaged to take the reindeer, without a stopover, directly from Norway to America. It was a reasonable solution to the problem of quarantine, and although it would restrict the field of choice for Lieutenant Devore in locating such a ship, and possibly more time would be required to find the ship, Jackson felt able to depart for the continent that evening, to carry funds in person to Kjellmann for purchase of the needed deer.

Before leaving London, he found ten minutes for a respite at St. Paul's Cathedral, and Max Jansen came to the Cecil Hotel to wish him well, but his words and manner only made Jackson feel uneasy again about Mocatta, Jansen and Company, which he no doubt discussed with Lieutenant Devore. Crossing the Channel aboard the steamship *Köningen Regentes*, he allowed himself to spend five dollars of official funds for a stateroom. On trains and ferries, via Hamburg, Kiel, Copenhagen, Helsingør, and Halsingborg, he arrived in Christiania (now Oslo) on 6 January 1898.

In the capital city, Jackson was welcomed at the American Consulate; Consul Henry Bordewich (from Minnesota) accompanied him to the Department of Interior Affairs of the Norwegian government, where the secretary provided important documents requesting local authorities to assist the Relief Mission. As well, Jackson learned that transfer of money by telegraph was commonly done in Norway, and while he would defer dealing with Kjellmann's needs until his arrival in Trondheim, he visited one of the banks to draw funds on his letter of credit; there, he was delighted when the manager gave him a card of introduction to staff at the university, providing admittance at no cost to the famous exhibit of the Viking ship. He had time also to tour the city in a hired sleigh (the fare must have been very little), and spent the evening at the home of the American consul, whose oldest son saw him through bitter cold of the streets at night safely back to the Victoria Hotel. Next day, at midnight, he arrived by train in Trondheim.

Some 150 kilometers southeast of that city, the town of Røros lies at the western edge of mountains rising to near 1600 meters in elevation, the slopes dense and rich in lichens of the kind (especially *Cladonia* spp. but perhaps also *Cetraria* spp.) favored as food by reindeer. In December 1897, on his way to Lapland, William Kjellmann had made a survey there, to ascertain the prospects of obtaining reindeer moss, and had informed the War Department that the farmers near Røros could provide without difficulty the amount of moss deemed necessary to sustain five hundred to six hundred reindeer on the journey across the Atlantic and North America. By the time Dr. Jackson had arrived in Norway, nine hundred loads of lichens had been drawn over the rough mountain roads by horses to the train station of the little town, and were being held there, awaiting payment. Jackson spent most of his day in Trondheim at the brokerage agency of E. A. Tonseth, settling the accounts of the farmers at Røros and confirming that the moss would be compacted into bags and sent as freight on the railway to the docks at Trondheim, ready for loading onto Lieutenant Devore's steamship.[50] After dispatching funds by telegram to Kjellmann in Alta, Finnmark, he was on his way north aboard the steamship *Vesteraalen*, bound for Tromsø; there he was transferred by rowboat to the steamer *Sigurd Jarl*, which was heading for Hammerfest in spite of the strong winds carrying sleet and snow and whipping the sea into powerful turbulence.[51]

## VII. Jackson in Finnmark and Devore in England

The storm that was sweeping in a gale over the fjords and mountains of Finnmark in January 1898 may have been one of the most severe recorded for a quarter of a century, since 1873, when the Meteorological Institute of Norway began countrywide observations.[52] When Sheldon Jackson arrived in Alta on 13 January on the little steamer *Nor* from Hammerfest, to anchor just offshore at the harbor at Bossekop, near-blizzard conditions had prevailed for a week over the whole of the northern Scandinavian Peninsula that makes up the province of Finnmark.[53] In early evening, in darkness and blowing snow, Jackson was rowed to the wharf in a small boat. There William Kjellmann had planned to meet him, but Kjellmann was not at the landing nor at the Alta Hotel. The hotel man's knowledge of English language exactly matched Jackson's understanding of Norwegian; they were entirely unable to communicate with each other, but the muted guest was shown to quarters all in readiness for him, a snug and comfortable room with featherbed of eiderdown and elegant sheets and coverlets of linen, hand embroidered. The hotel itself was not unattractive; it was a sturdy, two-story structure, in Nordic style of architecture, of logs on a thick, stone foundation, and it looked strong as a fortress. The next day, Jackson found that the telegraph operator and the postmaster knew a little English, so that he was easily able to report to Secretary of War Alger and later to contact Lieutenant Devore without the hoped-for assistance of Kjellmann. Kjellmann actually was on his way to Bossekop from Kautokeino, many kilometers to the south, but delayed by the storm and lost for a time in the mountains in the driving snow. He turned up at last three days later, exhausted, but with heartening results to report to Dr. Jackson: five hundred trained-to-harness reindeer, and all accoutrements for reindeer sleighing, were assured for purchase, and twenty-three men were willing to join the Relief Mission. A telegram from Lieutenant Devore reported further good news, that a ship had been chartered and would be able to leave from Scotland for Trondheim on 11 January.[54]

Exactly by what means Lieutenant Devore found his steamship is not known. Weeks after the mission in Norway was over, and the reindeer were in the United States, Devore received a letter from the Office of the Adjutant General, War Department, informing him that he had violated army regulations, paragraph 805, by not making reports to the adjutant general when he was in Europe with Dr. Jackson. In reply, Devore sent a copy of his orders, which were the instructions he had received on 23 December 1897 from the secretary of war pertinent to that special duty; Secretary Alger had not included any requirement that he report to anyone in the War Department. The lieutenant was a dedicated military man who knew well the regulations as they applied to his assignments, and he would follow his military orders exactly.[55] He sent few official reports (evidently all but one from Europe were cablegrams of one or two lines) to the secretary of war, and as a consequence, some of the details of his course of action in London are not known. One notation (18 March) indicated that he reserved the ship on 7 January and ensured its availability by agreeing to pay demurrage, notwithstanding the high rate demanded, so that it would be held in port (at Greenock, Scotland) until the appropriate date of departure for Norway. That date would be chosen in order that the material (lichens) at Trondheim be loaded on the ship in close coordination with the time the deer and personnel at Alta were assembled and ready to leave.

The vessel chartered was the *Manitoban* of the Allan Line of Glasgow and Montreal, a weary old veteran of the sea, built in 1865 by Laird Brothers of Birkenhead, England, as the *Ottawa*, and rechristened to her present name in 1872 after substantial alteration.[56] From Glasgow on 10 January, Devore found time to write a letter of three paragraphs to the secretary of war, giving some details about his tour of inspection of the ship. She was then being used as a cattle ship, just out of dry dock, with fresh paint and in satisfactory (actually, good, according to Devore) sanitary condition. With regard to the outbreak of disease in cattle in Sweden, she had not been connected with any commercial enterprise involving the Scandinavian countries and therefore was considered to be uncontaminated by infectious organisms that affect livestock. Allan Line's workmen were fitting her up to carry 500 reindeer and 150 passengers, and Devore made special request for a crew of experienced sailors so that the deer would be cared for and not suffer in case the attendants hired in Finnmark became seasick and unable to stand duty. He commented in his letter to Secretary Alger that Lapps were not accustomed to sailing—perhaps a pardonable generalization by Lieutenant Devore who did not know that many Saami are skilled seafarers. (As it turned out, on the forthcoming voyage many were affected by seasickness.)[57]

The *Manitoban* left Greenock on the Clyde on 15 January, for Trondheim, with a crew of fifty-three and a master, Andrew G. Braes, of long maritime experience; Captain Braes had been sailing the oceans for more than forty years.[58] The log books of the many journeys of the *Manitoban* evidently have not survived, and her route in January 1898 is not fully known; Captain Braes probably headed out of the Clyde for Norway in direct route, by way of the Sea of Hebrides and the Minch, sheltered perhaps from the force of main ocean, then went south of the Shetland Islands and across the North Sea to Bergen, where he took on board a Norwegian coastal pilot to guide his ship through Trondheimsfjord. The *Manitoban* arrived in Trondheim on 21 January, and Lieutenant Devore was waiting; he had arrived almost a week before. The expedition was already behind schedule.[59]

While he was in London, Devore had accepted some assistance from the Mocatta, Jansen Brokerage Company, as had Dr. Jackson, perhaps as a result of pressure from the brokers, but conceivably with reflection that in Washington, D.C., the Department of War (Secretary Alger and Assistant Secretary Meiklejohn) had indicated trust in the company, and with recognition that Jansen and Mocatta as commodity brokers had considerable knowledge about reindeer in Norway, where also they had business contacts.[60] Devore was concerned, as well, especially after Jackson had left for the continent, about Kjellmann's results in Finnmark; that the five hundred reindeer would be available there was hoped for but not certain, and no doubt Devore thought that having alternative sources would not be unwise. In addition, Devore felt bound to meet the schedule given by Secretary Alger, that the deer be en route to America by 20 January. Kjellmann, too, had received instructions from the secretary by cablegram on 24 December 1897 that the animals must be shipped by 20 January (instead of 15 February as Kjellmann had indicated by cablegram to Alger on 22 December). Thus, when B. Mocatta and Max Jansen drew up a document, dated 4 January 1898, Lieutenant Devore signed the paper, thereby agreeing that the Mocatta, Jansen Company would act as agents to assist the Yukon Relief Mission in chartering a steamer and obtaining in Norway bids for provision of five hundred deer and the necessary lichen fodder. Devore viewed the document as an agreement whereby the Company was permitted to assist in obtaining information, a service for which compensation was to be ten percent of cost, payable at the port of embarkation. Jansen and Mocatta evidently interpreted the paper as an actual contract, giving them privilege to procure a ship and purchase up to five hundred reindeer and a quantity of reindeer moss, to be charged to the account of the Department of War, at cost plus a fee. Inherent in the document of 4 January was mutual misunderstanding.[61]

Less than a week went by before Devore, in his negotiations with the Allan Line for the *Manitoban*, began to sense that the Mocatta, Jansen brokerage firm meant to do much more than advise. As concerned finding a ship, the company had determined location of available vessels and then perhaps importuned to be considered the intermediary agency, for a fee, but Devore declined the firm's proposals, which possibly stirred Mocatta or Jansen into making unflattering comments. In his one report of any length (as noted, three paragraphs on 10 January) to Secretary Alger, Devore said that the company was giving "lots of trouble," but in what ways he did not describe.[62]

Just before leaving London for Glasgow and the Allan Line's headquarters there, Devore received from William Kjellmann in Alta a telegram stating that the deer could not be shipped from Alta (Bossekop) until 15 February. Devore's apprehensions about Kjellmann's results were thus confirmed, but as well, Kjellmann in effect was making a new schedule, postponing time of departure more than three weeks, a change of grave concern inasmuch as the secretary of war had emphasized repeatedly to his military aide as well as to Dr. Jackson (and at least one time to Kjellmann) that the animals must be en route to New York by 20 January, for any delay might lead to disaster for the people in the Yukon District. Thinking of the emergency amongst the miners, of Alger's orders to the leaders of that military mission, and probably of the great cost of holding a steamship in port until the fifteenth of February, Lieutenant Devore must have been highly troubled after reading that telegram, especially as he was unable to contact Jackson, who had departed from Trondheim on 8 January and was somewhere at sea along the coast of Norway, en route to Finnmark. As became evident later, Devore discussed the situation with Mocatta and Jansen, and he decided (by the next day, probably after thinking most of the night) to permit the brokerage firm to attempt to find as many as five hundred deer, the quantity accepted to be dependent on the number that Kjellmann might have purchased, in time to load onto a vessel by 22 January, which effort if successful would be compensated by a payment of three thousand dollars and an extra one hundred dollars for each day gained. All of that he put in writing, and indeed it read like a contract, with firm specifications. Verbally, he informed the two brokers that he was not authorized to pay for anything without approval of Dr. Jackson, and he instructed Jansen (who was ready on the instant to go to Norway and begin to collect deer and lichens) to communicate with Jackson in Alta on or about 13 January and make full report of any negotiations.[63] Devore considered his written instructions to Mocatta and Jansen somewhat as an addendum to the document of 4 January, and neither paper was to be interpreted as a definite contract permitting either purchase or payment, but by issuing them he was trying to cover all contingencies; i.e., if few deer were obtained in Finnmark by Kjellmann, then possibly the desired complement could be made up through those sought out by Mocatta and Jansen, in time for shipment as ordered by the secretary of war. Jansen departed at once for Norway, contacted his commercial colleague, the commodities broker S. M. Sivertsen in the town of Mosjøen (on Vefsnfjord, some 300 kilometers north of Trondheim), and ordered five hundred reindeer and a large quantity of lichen.

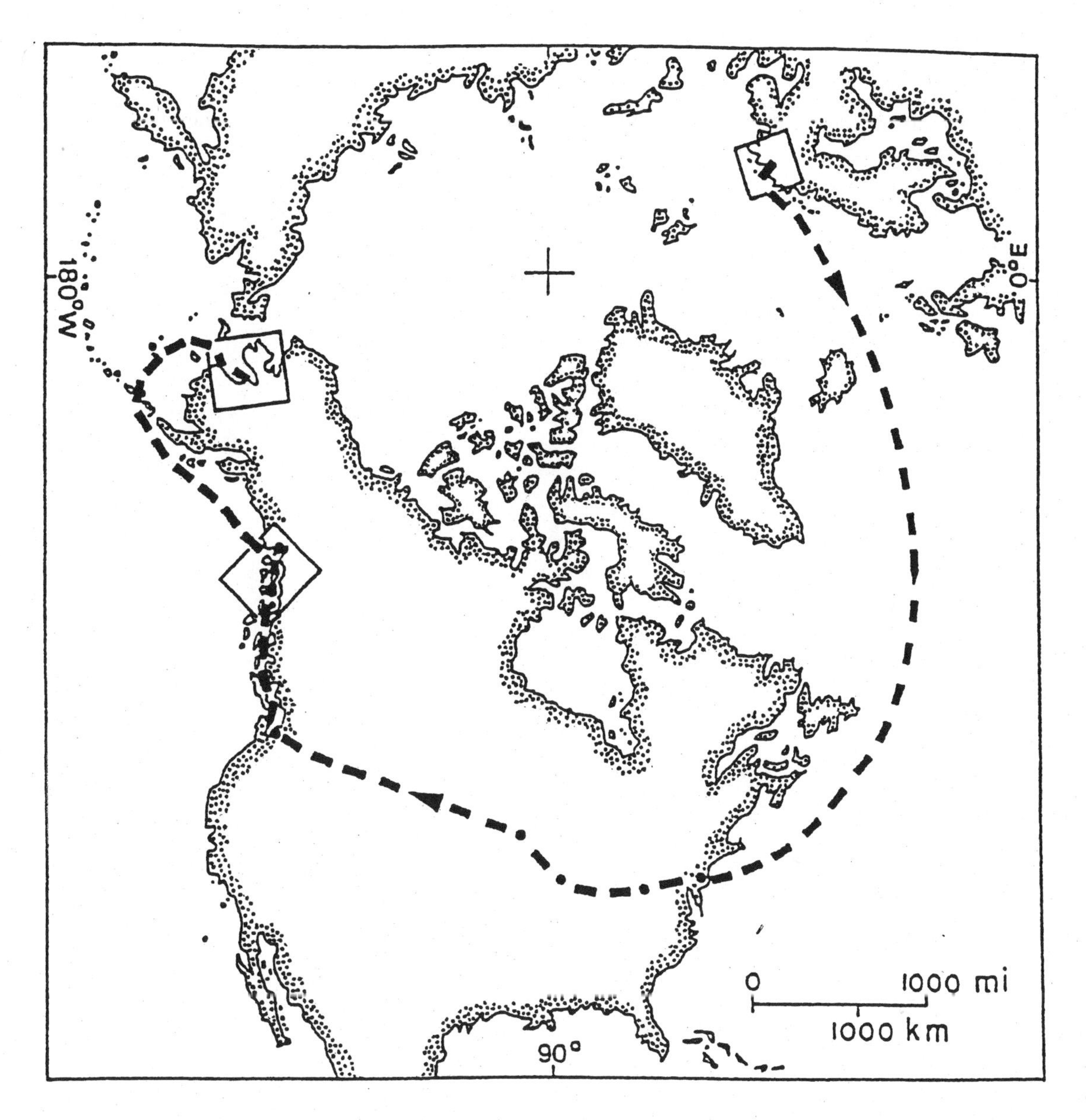


The route of the Yukon Relief Expedition. The small rectangles show areas enlarged (maps, pages 39, 95, and 178).

Devore had no way to know that the information that Kjellmann had sent by telegram was out of date, wholly passé, by 11 January. Kjellmann by then was in Kautokeino and had despatched emissaries to every likely settlement in central Finnmark and one in Finland; they found that reindeer were not at all in short supply and that herd owners were eager to sell, so that on that date, and perhaps earlier, the buyers had obtained more than the desired number of deer, and they had been able to hire some reindeer managers.[64] No one sent that information to Devore, who had only Kjellmann's telegram; thinking that by his action in enlisting the help of Mocatta and Jansen the problem of lack of reindeer might be solved, Devore wrote out a message on 11 January and sent it via telegraph addressed to Jackson in Hammerfest, apprising him that according to Jansen all components (deer, moss, and necessary gear) could be aboard ship by 22 January. The message probably was not received by Jackson; Devore in Trondheim, trying to understand why the reindeer moss was not in the sacks or bales that Jackson had requested and had paid the Tonseth brokerage to arrange, and attempting to set matters right, had no word from his colleague until 17 January.[65]

Reindeer moss filling seventy-one railroad cars was at the port to be stowed onto the *Manitoban*. Delayed by storms but with the best coastal pilot from Bergen aboard, the ship arrived at Trondheim on 21 January, but because her draft of twenty-one feet exceeded the limit of the harbor basin, she had to anchor out a little distance from the quay.[66] She was there a full week while the reindeer moss was being loaded aboard, laboriously transferred once from railroad cars to barge and a second time from barge to the ship's hold, in snow and winds whipping the waters of the fjord into a tumult of waves so rough and dangerous that the work had to be stopped lest the barge founder in the sea and lives be lost. Whether none of the moss had been packaged is not clear; Devore stated that the moss was broken up and not sacked, but even sacks or bales would have been difficult to move about in the conditions of weather that week.[67] A large quantity of lichen was left behind at the dock when the ship departed; the Tonseth Brokerage Company gave free moss at anyone's request and during the following days sent out many loads of it for use on the nearby farms.[68]

The *Manitoban* made her way westward out of Trondheimsfjord on 28 January and northward along the coast, and she arrived in Bossekop harbor at Alta on 2 February, the entire time in strong winds and a heavy sea. Lieutenant Devore was the only passenger. In a sense, she was his ship, but he probably found little pleasure in the trip to Alta; he had just been informed in Trondheim that legal action was being brought against him in Norway for breach of contract.

## VIII. Finnmark, January and February 1898

At Bossekop the weather had not changed since early January; sometimes the winds subsided, but only briefly, and the snow continued to fall and pile up in deep drifts. That such conditions were not of much concern to the local people rather astonished Sheldon Jackson. Kjellmann, too, would not be distracted by the storms in his plans to travel to the south to complete the transactions for the deer and drivers. Jackson remained at the hotel, always on duty, but he found time to meet some of the people and to attend services at the local church. The church building was icy cold within, and he could not understand a word of what was going on, but he sat amongst the congregation for the three Sundays spent in Alta (on the last one Kjellmann went with him), and observed the ceremony.[69]

He was in contact again with Lieutenant Devore. Devore's disbursement (from Trondheim) of funds to Jackson as necessary had gone smoothly via the telegraph in Norway; Jackson took the opportunity during this relatively quiet interval (i.e., before Kjellmann and his men returned with reindeer and Lieutenant Devore arrived with the *Manitoban* ) to make a careful tabulation of expenses to date (one item showed that he had paid ten cents for help of an interpreter; Jackson was extremely cautious about spending federal money). On 19 January, he bought reindeer moss (about sixty tons, baled, and delivered at Bossekop) through the commodities broker at Mosjøen (S. M. Sivertsen), who he said with alarm "was buying against us in this market, runs the price up from 10 to 17 kroner per load." Two days hence, Sivertsen by wire informed him that the five hundred reindeer ordered (through Sivertsen's brokerage company by Max Jansen of Mocatta, Jansen and Company) for the Yukon Relief Expedition were arriving from Sweden on 24 January at Skibotn (on Lyngen Fjord, east of Tromsø, not far distant from the Norwegian-Swedish border), and he wished Dr. Jackson to tell him what to do with them and when Jackson would come and get them. Evidently, Max Jansen had never made clear to Jackson the details of that order nor had he ever received approval from Jackson for the order; inasmuch as Jackson had not expected any reindeer in addition to the five hundred already acquired in Finnmark, he told Sivertsen that he would not receive those deer, at Skibotn or anywhere else. Sivertsen's deer were available somewhat later than Max Jansen had predicted to Lieutenant Devore, but actually they might have been loaded on ship more expeditiously than those being driven to Alta in the immoderate weather, and many kilometers yet to the south.[70] Jackson's refusal made Sivertsen more than upset; he was angry. His message of 21 January was probably the third or fourth he had sent to Jackson within as many days.

The storm developed to near hurricane strength during the last week of January as the herds were headed for Alta and

Bossekop. From Bautajok, Karasjok, Kautokeino, and Masi in Norway, and Inari in Finland, far to the southeast, with drivers attending, more than 550 reindeer began the northward trek, accompanied by the men and women and children who had signed up to go to America. With the families and their relatives bundled snug and warm in pulken, some aligned in tandem hitch, ten pulken to ten reindeer, the trains slowly made their way overland through the wind and snow. Dr. Jackson in Alta felt the hotel building shake in the pounding wind; the staff of the hotel came round to nail blankets over every window in his room lest the panes of glass could not withstand the blows. The outdoor temperature was around -10 degrees C (+14 degrees F), not an uncomfortable ambiance, but the chill of the wind must have produced an extreme cold. Jackson offered prayers for the protection of the people coming through the storm, though he (and the mayor of Alta, who came to call on him during the day) doubted that anyone, even the Laplanders, could travel during that time. Snow in deep drifts blocked all the lanes in Bossekop and Alta. The steamer from Hammerfest could not get through Altafjorden; besides other things, it was carrying precious mail. On 27 January, Jackson did not step outside of the hotel, probably the sole day he missed going out for a walk about the town for exercise and fresh air and stopping by to visit new friends.

When William Kjellmann arrived at approximately half past noon at the Alta Hotel on 28 January, Jackson must have felt a sense of incredulity, as well as joy and relief, not only to see his much-trusted friend but to hear from him that travel was possible in spite of the weather; later that afternoon, after four days en route, Mathis Eira and his men came from Masi with ninety deer. In fact, the weather seemed to be improving slightly. Children were out of doors again, running about the settlement and nearby slopes on their skis. Their activity so much impressed Dr. Jackson that he commented, "They are very expert. [Skis] are a great improvement upon the snowshoe of the American."[71] From Bautajok, Samuel Kemi and Carl Suhr led a group with 114 deer. Kjeldsberg and Rist and a company of forty-four brought in 252 deer from Kautokeino. The fourth and last herd of ninety came in on 1 February with the merchant Paulsen and his men from Karasjok.

That night, the *Manitoban* slipped into Bossekop harbor, cast anchor, and stood to shore a short distance out from the wharf. Preparations for departure began early on 2 February. Jackson gave to contractors in Bossekop the job of loading the deer aboard ship. Using three large rowboats that went from shore to ship, the workmen ferried the deer across to a large, platformed barge they had secured alongside the *Manitoban*, and encouraged the animals to walk by themselves up the gangplank to the decks. All to be taken to America (539) were installed on the vessel by 3 February, along with 418 sledges and 511 harnesses.[72]

Some of the personnel who intended to go with the ship evidently had documents yet to sign, so that the process of boarding was not accomplished quickly but went on throughout the afternoon and evening of 3 February. Jackson provided only a few details about the agreement negotiated in January 1898 between the Department of War and those going as departmental employees to Alaska, but the contract in full was published in his report to the U.S. Senate (through the secretary of the interior) in 1898 (see Jackson, 1898b). Some individuals put their signatures to two versions of the agreement, one as that published and another in abbreviated format, given here in entirety:[73]

## Contract of Service

All the undersigned admit and make known by these presents to have hired themselves as reindeer drivers in Alaska and to execute furthermore such work as our superiors appointed by the United States government may require; Also to look out for and take care of reindeer under the transport to Alaska. Even so we bind ourselves to behave our selves well and decently and show discipline. For services above referred to undersigned Sheldon Jackson binds himself on behalf of the United States government to procure the payment to every one of the undersigned of a yearly salary of Kr. 1000 (one thousand crowns) with good and sufficient food and clothes (to bacco not included) and free transport to Alaska; furthermore shall the hired men have free nursing and medicine and also the salary during time of sickness.

The time of notice of end of service is put mutually to six (6) months. This contract enters into effect from and on the lst. day of February, this year.

Bossekop, February lst., 1898.
[Signed] Sheldon Jackson, Special Agent
United States War Department.

The language of the short form is irregular and somewhat awkward. It may have been written up in a hurry, and signatures were taken on it *pro tempore* from those men who arrived in

Bossekop with so little time remaining before the *Manitoban* was to leave that strict formality could not be observed, for the last day in port in Finnmark was almost a confusion of activity for the overworked responsible leaders, of whom there were only three, Jackson, Devore, and Kjellmann. Among those coming in late was Carl Johan Sakariassen, who during the recent storm had walked and skied over the mountains from Fladstrand to Talvik; his signature was on both versions of the contract.[74] The names of all who joined the relief expedition were published as well in Jackson's report. Those lists have some errors in spelling of names and in designating provenance of personnel, and one or two may be listed twice. Most of the people were from Kautokeino. In Church Register No. 5 (1896–1916), for Kautokeino Parish, fifty people (including wives and children) are enrolled as leaving the parish, bound for Alaska; the dates of departure, from 21 to 30 January 1898, given for each in the register probably indicate the approximate time of signing contracts with the War Department of the United States.[75] Isak Johannesen Haetta, numbered thirteenth among the emigrants on the register, kept his document for many years; his grandson, Jan Henry Keskitalo, permitted its publication by Professor Hugh Beach in 1986. Isak Johannesen Haetta's contract was dated 13 January 1898.[76]

Sixty-eight men signed formal agreements with the United States Department of War. Four women (Sofie Andersen and Marit Biti from Karasjok, and Berit Nilsdatter Eira and Ida Johannesdatter Haetta from Kautokeino) also signed contracts (identical in language to those signed by men).[77] The total number of people going to Alaska was 113 (according to Jackson's report). The newspaper *Finnmarksposten* (Hammerfest) on 15 February 1898 published the contract as presented in the Norwegian language, and a list of those who were to travel to North America.[78]

For many who had signed contracts of work for the Department of War and would depart from families, friends, and homeland, the third day of February 1898 was a time of distress, although their acceptance of a period of work in Alaska was not generally the result of a hasty or even recent decision. Kjellmann had been traveling about in Finnmark for almost two months, and Mathis Eira, Samuel Johnsen Kemi, Aslak L. Somby, and Per Aslaksen Rist, the Saami who returned home to Finnmark from Alaska in December 1897, were well known in the province. Word about them and Jackson and Kjellmann no doubt had been passed from one person to another for weeks, in church and visits among friends, so that every community in northern Norway had opportunity to hear about the Klondike/Yukon Relief Mission to help the "gold-diggers," which was the term for the miners used in some of the Norwegian journals; *Finnmarksposten* (Hammerfest), and *Dagsposten* and *Trondhjems Adresseavis* (Trondheim), and other newspapers had given news quite regularly about the project. In the far north, the first account to appear was in *Finnmarksposten*, 21 December 1897, reporting that Kjellmann had come from America to buy one thousand reindeer and engage one hundred men to be sent to the "Goldland Klondyke," and that individuals hired some years earlier had that month returned from Alaska to Finnmark, each with three thousand kroner and a good lot of experience.[79] Thus, those who had chosen in January 1898 to accept work in Alaska had some weeks to think about their contracts, and about departure. Nonetheless, on the third day of February, actually and finally to leave was very difficult. Perhaps on that account some of the Lapps (as the Saami then usually were called) were drinking alcohol (whiskey, Jackson said) and decidedly feeling its effects, stalling to impede departure, demanding time to make an inventory of all their personal goods, declaring intent not to go to North America, so adamant in attitude that the chief of police threatened to throw them all in jail.[80] Then the steamer from Hammerfest arrived, no doubt to the joy of those hoping for letters and needed items, which had come just in time to be put aboard the *Manitoban*. Some of the Saami were especially delighted, for on the rowboat as it approached the shore, amongst the mail and freight from the steamer were the four kegs of whiskey they had ordered from Hammerfest by telegraph, intended to be stowed aboard for the journey. Kjellmann had been forewarned, and the men could only watch as the kegs were carted away by the chief of police, but it made them very unhappy.

Gradually, as calmness again prevailed, the departing men, women, and children went to their quarters on the *Manitoban*. Their living space on the ship turned out to be open steerage, below the main deck, where they were situated almost as were the reindeer. To be traveling thus, as employees sought out by the Department of War of the United States of America and by the nation's Congress, must have been bewildering, and added to the trauma of farewells.

At 2100 hours on 3 February, Dr. Jackson came aboard. He was tired out; the details, social and otherwise, attendant to departure had been many, and in addition to those, he and Lieutenant Devore undoubtedly had spent hours discussing the wretched legal problems that involved the expedition, or at least that implicated Devore, as one of its responsible leaders. Jackson mentioned nothing of that in any records, but the success of the expedition may

Opposite: Route of the *Manitoban* out of Altafjorden, in Norway. The dots coursing to the south of Alta indicate the direction of localities providing reindeer for the expedition.

have seemed at that time anything but triumphant, especially since he had been deficient in some of his own responsibilities, which he must have hoped would escape public attention. All day on 3 February, he had been in a torment of pain in both his knees, from "rheumatism," so that walking down the steep slope to the wharf, as was necessary several times in morning and afternoon, had been like torture. In the evening a hired sleigh carried him to the lighter. Shipboard, his room was cold; steam had condensed over everything. The bedding on his berth was wet. He could not sleep. Some of the Saami did manage after all to obtain a few bottles of whiskey and Jackson said that they "made the steerage a pandemonium all night."[81] From the telegraph station in Alta, before his sleigh ride to the quay, Jackson had sent word in a cablegram to the Department of War that "530 reindeer and 87 Lapps" (adults) were about to leave for New York.

Anchor was weighed shortly after Kjellmann came aboard at about 0400 hours, 4 February, and the *Manitoban* sailed once again on Altafjorden. Two Norwegian pilots guided her passage northwestward along the coast through Stjernsund, into Lopphavet, then southward a short way (approximately twenty kilometers) to the tiny island of Loppa (Loppen), and into the harbor of the island's small settlement. The *Manitoban* halted by the landing only long enough to drop off the two coastal pilots, the Norwegian customs officer, and Lieutenant Devore, all of whom intended to go to Tromsø on the local mail steamer that called at Loppa regularly. From Tromsø, Devore was going south, back to Trondheim.[82]

## IX. The Courts of Justice in Christiania

Devore's abrupt departure on 4 February from the *Manitoban* at Loppa was not fully explained in official documents. To the War Department (and to Congress), Jackson reported that en route to the United States from Bossekop, Lieutenant Devore traveled to Trondheim and London; in his personal journal he wrote that Devore went via London to resolve matters, and because the Mocatta and Jansen Brokerage Company had made trouble, Devore was to carry along certain original communications (telegrams) that the company and its agents had sent to Jackson at

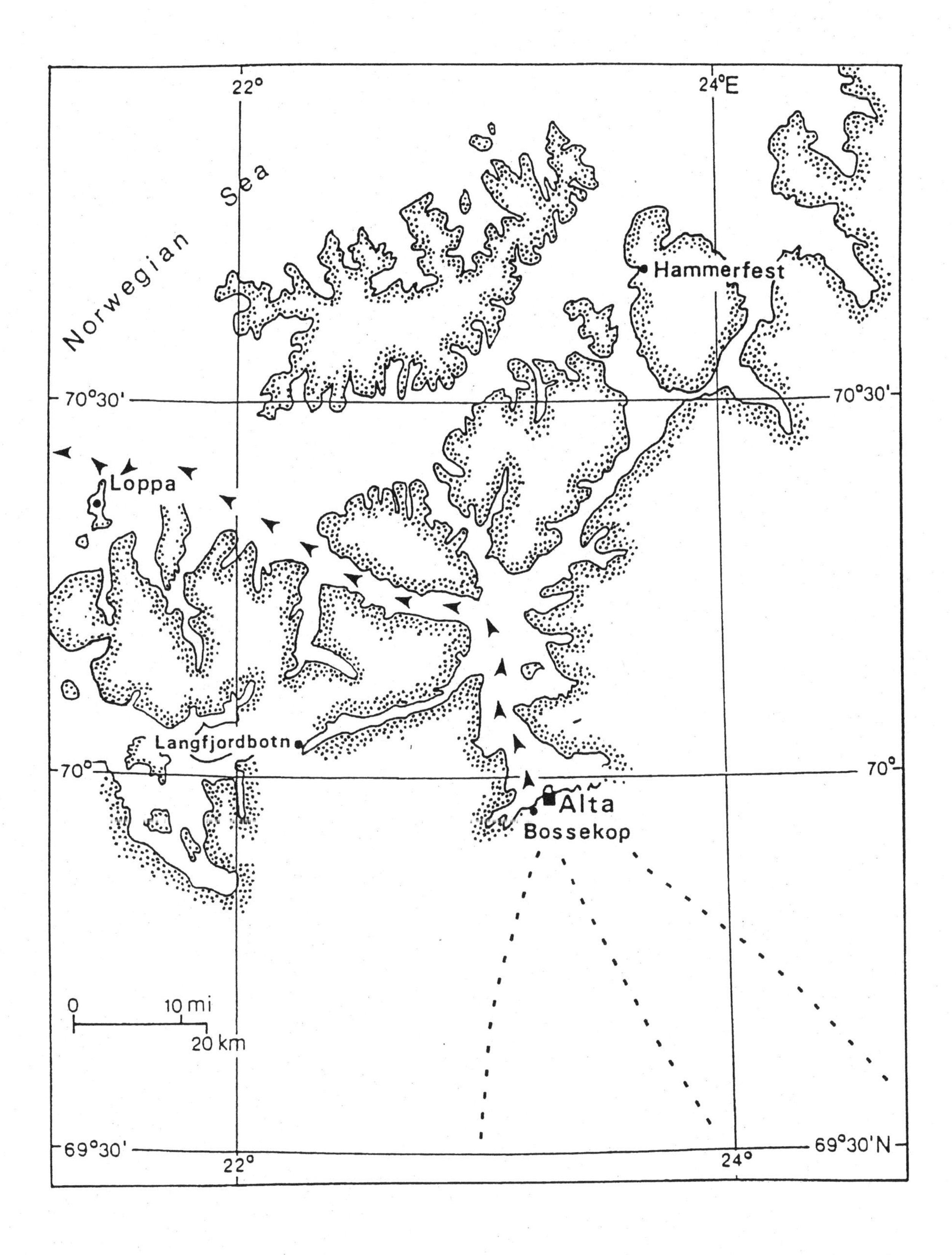
22°
24°E
Norwegian Sea
Hammerfest
70°30'
70°30'
Loppa
Langfjordbotn
70°
70°
Alta
Bossekop
0
10 mi
20 km
69°30'
22°
24°
69°30'N

Alta.[83] The trouble, not defined, would involve litigation, and Devore was not able to travel with the relief expedition to New York on the *Manitoban* as planned, but instead had to remain in Norway because his presence again in Trondheim was demanded by Norwegian police commissioners. Jackson evidently did not make written report about the matter to his superiors in the War Department.

The legal papers were served in January. Devore was in Trondheim, supervising loading of reindeer moss to the *Manitoban* during the great storm. While the work was going on, Devore was summoned to appear at the judicial court, accused of committing a felony as defined in the penal code of Norway. If the papers had been delivered to him at the last minute before the *Manitoban* sailed for Bossekop, he would have had no opportunity to compose and send a message to Alta, but more likely the summons was served well before departure, and he had ample time to let Dr. Jackson know about that latest, most unpleasant development. Among the telegrams preserved in archives and published, dating from the expedition's last few days in Finnmark, one from Jackson to Devore indicated that Jackson knew of the impending litigation by 22 January 1898. On that day he advised Devore in Trondheim immediately to inform the American consul there and ask for help concerning Max Jansen before leaving Trondheim.[84] A small consular unit might have been maintained in Trondheim, but the consul and his offices, i.e., the American consulate in Norway, were in Christiania, and the official to whom Lieutenant Devore appealed was Henry Bordewich, who had met and assisted Sheldon Jackson the month before. But still the record as published by Jackson was not correct, for the litigation arose not from Jansen, but from S. M. Sivertsen, the livestock broker at Mosjøen. Sivertsen with Lensmann Peter Rivertz of Troms Province had made formal complaint to Norwegian authorities against both Daniel Devore and Max Jansen, who they charged had placed an order for five hundred reindeer and associated gear. As of about 24 January, the deer, as well as harnesses, lines, and pulken, and some reindeer moss as well, purchased in Sweden and Norway by Rivertz at the request of Sivertsen, were to be at the Port of Skibotn, ready to be sent to North America for the Yukon Relief Expedition; that information Sivertsen sent by telegram to the leader of the expedition, Dr. Jackson at Alta, who evidently was astonished to receive the message, for Kjellmann in southern and central Finnmark had already bought some five hundred reindeer to fulfill the number authorized by the War Department. Jackson not only refused to accept Sivertsen's animals, but he declared to

Sivertsen that he knew nothing about them. Perhaps in compliance with merchant law, S. M. Sivertsen sent several messages to Dr. Jackson, asking him to accept the reindeer. Jackson either refused or did not reply, and Sivertsen and Rivertz, after consulting their attorney (Wulff), appealed to the courts of Norway. In the lawsuits that developed, Sheldon Jackson was not named as a principal, for in fact he had not ordered reindeer from Sivertsen; Max Jansen had placed the order under his own interpretation of authority given in Devore's requisitions of early January.[85]

In Skibotn in the meantime, the five hundred reindeer arrived. No corral or livestock yard existed in the little town. Rivertz had to act quickly to keep his investment intact; for ten days, eight guards were stationed out to watch over the herd, and a fence was put up, to which the deer did not object so long as they had reindeer moss. But the supply of lichen fodder was not great, and as soon as it was depleted, the deer broke down a gate and scattered off into the mountains to find food for themselves. Only by sheer luck could the animals have been recaptured.[86] Evidently, efforts were made to do so, probably at about the time Lieutenant Devore took leave of the *Manitoban* to return to Trondheim.

From Loppa to Tromsø and south to Trondheim was a journey of two to four days, depending on season and weather. On or about 7 February, when Lieutenant Devore (perhaps under escort) walked off the little coastal steamer as it arrived in Trondheim from the north, officers of the police department were waiting; he was taken into custody, not thrown into jail, but forbidden to leave his hotel. Max Jansen also was under arrest and restricted to quarters. As the cases (there were two suits, initiated respectively by the two plaintiffs, Rivertz and Sivertsen) were to be heard in judicial court in Christiania, Jansen and Devore were taken to that city. There the American consul and the attorney (V. M. Aubert) serving the consulate were present to advise Lieutenant Devore. The Mocatta and Jansen Brokerage Company retained the London solicitors Emanuel, Round, and Nathan to assist in the case, and in the courtroom, Max Jansen was represented by attorneys Schjodt and Voss. The plaintiffs, Rivertz and Sivertsen, were represented by attorney Arndt Schiander, who was qualified to practice before the Supreme Court of Norway. In Christiania as in Trondheim, Devore and Jansen were forbidden to leave their living quarters. Devore's place of residence was not recorded; Jansen was staying at the Hotel Scandinavio.[87]

The two cases would be discussed in separate courts; Case No. 212, plaintiff Peter Rivertz, went to the municipal court, and Case No. 213, plaintiff S. M. Sivertsen, was heard before the judges of the Gjesteretssag (the Guest Court, which attended legal actions involving citizens of foreign countries), but as pointed out by the judges, the cases were identical except for person of plaintiff (and reparations sought). An examination of the charges apparently indicated that the financial losses of Rivertz as well as of Sivertsen could be assigned to fraudulent contract given by Jansen with attribution to Devore, both of whom if found guilty could be punished by fines and/or by imprisonment up to three years, as defined in the penal code for such a felony. The cases were to be judged by the courts; seven lawyers, four and possibly six judges, the consul of the United States and some of his staff, and no doubt a number of assistants and clerks of law were associated with Case 212 and Case 213 in February 1898, and over the months to follow, others gave affidavits and testimony.[88]

The first hearing on the lawsuits took place on 10 February 1898. Accompanied by Consul Bordewich and attorney V. M. Aubert, Devore appeared before the court and to the charges presented against him, he explained through counsel that as disbursement officer for the Yukon Relief Expedition, he had not been given authority by the secretary of war of the United States of America to instruct anyone to buy anything for the Department of War; he did not order the reindeer. The statements were verifiably true, but he was asked to give reason for the two documents he had signed in London on 4 January and 8 January with the brokers Mocatta and Jansen, which in essence represented the source of all the problems inasmuch as they appeared to be legal contracts, permitting Mocatta and Jansen to purchase items for the Department of War, and one of the papers indicated precisely the payment to be made to Mocatta and Jansen. Having reference to those papers, Max Jansen had asked the Sivertsen Brokerage to obtain five hundred reindeer for the Yukon Relief Expedition.[89]

Concerning his two days (10 and 11 February 1898) in court, Devore later reported orally to the War Department, but in advance of that time, the attorneys of his codefendant Max Jansen had taken care to write to the department about the litigation. Emanuel, Round, and Nathan on 21 January had sent copies of the two documents (of 4 and 8 January) to the secretary of war, and Alger had turned them over to his adjutant general for advice. Two months later (5 March 1898), Adjutant General H. C. Corbin ordered Devore (in the tempered language that was to be accorded the military aide of the secretary of war) to give immediate explanation concerning the two incriminating pieces, to which command Devore stated in writing to the adjutant general (as the court in Norway had been informed) that the paper of 4 January gave permission to Jansen merely to obtain offer for as many as five hundred reindeer, whereas the note of 8 January gave Jansen license only to find satisfactory reindeer as had been described by Devore, those animals to be added to the number that William Kjellmann might have obtained in order to make up a total of five hundred deer, and all were to be on board a transport vessel by 22 January. Lieutenant Devore was following as closely as he could the orders given to him (and to Jackson and Kjellmann) by the secretary of war.[90]

The judges of the court did not immediately accept that construction. They held that Devore could be liable for plaintiffs' expenses, but whether he was and how much he might owe the plaintiffs were not at once apparent and would have to be found by the court through study of the two documents and examination of all parties to determine the verity of their claims. On 10 February, the summation of Rivertz' financial expenditures (itemized: reindeer; sledges; moss; harnesses; wages for watchmen, fence builders, and men from Lapland who were to accompany the deer to America; attorney Wulff's fees; expense for telegrams; and travel to court in Christiania) amounted to 24,506.35 kroner ($6624.06 U.S.). Rivertz demanded payment in full from Devore or Jansen, plus interest and court expenses. Sivertsen's claims amounted to 3293.20 kroner ($890.15 U.S.) (for brokerage fees, travel and communications expenses, fees of attorney Wulff, and purchase of moss).[91]

The process to determine the extent of liability of Devore could take many months, and the plaintiffs could appeal a judgment to a higher court; thus, conceivably, Lieutenant Devore faced long detention in Christiania, but attorney Aubert requested the court to permit Devore's absence from Norway during that time in order that he could return to duty in the United States as soon as possible. To that request, plaintiffs' attorney Schiander consented, but only conditionally; Devore must deposit in one of the financial institutions in Christiania eighteen hundred pounds (British) as surety. Those funds were to be left in trust in the name of the American consul, whose signature was to be on an attached document describing their purpose, and the money was not to be withdrawn except on approval of the court and with the signature again of the American consul. Attorney Aubert on behalf of Consul Bordewich and Lieutenant Devore agreed to the terms. After fulfilling the court's requirements, Lieutenant Devore would be free to go. Next day, eighteen hundred pounds were deposited in the bank of Thomas Joh. Heftye and Son in Christiania, Consul Bordewich unhappily signed his name to guarantee the fund, and Lieutenant Devore fled from Norway; he must have boarded a steamer that same day, for on 13 February from Queenstown

(Ireland) he sent a cablegram to the secretary of war: "Sailing *Campania*."[92]

Secretary of War Russell A. Alger was recovering from severe illness, but he was at his desk to hear Lieutenant Devore's report on 20 February, a few hours after the *Campania* docked in New York City. That day as well, the secretary received a cablegram from Christiania, asking that he instruct the consulate in Norway to understand that Devore's deposit of funds was also a surety for Max Jansen. The sender's name is not on record, nor is any direct reply. Next day, Acting Secretary of War Meiklejohn requested the secretary of state to advise the consul of the United States in Christiania by cablegram that the War Department was assuming the responsibility of Lieutenant Devore in connection with his recent duty in Europe, and that "the interest of the Government in any suits commenced against Lieutenant Devore or the United States by the Mocatta-Janson [*sic*] Company...should be carefully protected...and that the War Department assumes no responsibility for Max Janson [*sic*]..."[93]

When Lieutenant Devore departed from Christiania after his appearance at the Court of Justice, the Yukon Relief Expedition in Norway essentially had been completed, and successfully so (although far behind schedule), in spite of the unfortunate legal involvement. The one person who felt further responsibility there concerning the expedition was the American consul, in whose trust had been left American funds to the amount of approximately eight thousand dollars (eighteen hundred pounds British) as surety. Consul Bordewich was not pleased with that arrangement, and on 10 March 1898 he sent an official communication to Lieutenant Devore (then on duty in Washington, D.C.), suggesting that the state department order the consul in Christiania (that is, Bordewich himself) to regain for the United States the funds on deposit with Heftye and Son, Bankers. Devore took action and the order to the consul (in language of a request) came through official channels in early April. Bordewich referred the matter to the attorney for the consulate, who asked the court and the plaintiffs' attorney, Arndt Schiander, to consent to withdrawal of the surety funds on deposit. Schiander (and his clients) found the request acceptable, but again only conditionally, and the court agreed; before withdrawing the eighteen hundred pounds, the United States War Department must make an official declaration to the plaintiffs' attorney that the department assumed the suit and Devore's responsibility therein. The War Department complied by letter on 9 May 1898. The consulate's attorney, V. M. Aubert, recommended that £75 be retained in Christiania, in the account of the U.S. Department of State and available to the consul for payment of fees and miscellaneous expenses arising from the suits. Consul Bordewich sent the balance (£1725) to the London bank of Seligman Brothers on 17 June 1898, to be credited to the account of the Department of War and subject to the draft of Lieutenant Devore.[94]

In Christiania, Max Jansen spent almost four months under house arrest; in February 1898 he had deposited five thousand kroner ($1351.50 U.S.) with the Christiania police department and thereafter was free to step out of his hotel, but that amount, less than one-quarter of his alleged debt to Sivertsen and Rivertz, was merely to guarantee his presence in Norway. On 28 March 1898, Rivertz (probably through his attorney) legally impounded the money to prevent its removal from Norway, and Jansen himself was forbidden to leave the country. Consul Bordewich was well aware of Jansen's detention and told his chiefs at the Department of State that the suits conceivably could be carried to the Supreme Court and that if "Devore had not made the money deposit he would have been in the same fix [as Jansen]...." Around the middle of May 1898, Jansen was able adequately to secure his debt and was permitted to return to England. That the commodity brokerage business he owned with Mocatta did not survive Jansen attributed to the loss of his credibility in Norway and Sweden, where chiefly the firm had dealt. He eventually moved to Canada but continued to apply for redress, in communications (some were angry and some were threatening, since he felt that the War Department had been unjust) to President McKinley, the secretary of war, and the British ambassador to the United States Government, who was then Sir Julian Pauncefote. On 10 October 1898, Ambassador Pauncefote from his home in America at New London, Connecticut, wrote to the American secretary of state, giving his appraisal of the case of Max Jansen, with the conclusion that since Jansen appeared to have acted in good faith in the interest of the United States, to award some compensation for his losses might be appropriate.[95]

John Hay, the newly appointed secretary of state, had been serving his country as ambassador to Britain when Dr. Jackson and Lieutenant Devore had called at the American embassy in London in January 1898. He and Pauncefote were friends, and colleagues as well, who had worked together in efforts to solve other international problems. Upon receipt of Pauncefote's communication, Secretary of State Hay requested Secretary Alger to supply the State Department with copies of all documents pertinent to the appeal of Max Jansen, for the purpose of critical examination to determine responsibility. The Department of War was fully cooperative, copies were sent, and the offices of the British ambassador and the United States Department of State jointly conducted an investigation of the case. By letter dated 18 April 1899, Max Jansen was informed (by Assistant Secretary of War G. D. Meiklejohn) of the review, out of which was concluded that the Mocatta and Jansen Foreign Produce Brokerage Company was "fully if not actually over paid for all services rendered."[96]

In Norway, the cases of *Rivertz and Sivertsen v. Devore and Jansen* remained unsettled for almost two years while the authorities took testimony and reviewed all aspects of the lawsuits. Finally, on 27 January 1900, decisions were announced (by Municipal Court Judges K. Qvigstad, F. Prydz, and Jonas Stenersen). The court found that the agreements of 4 January and 8 January 1898, between Lieutenant Devore and Mocatta and Jansen, were not contracts permitting purchase; those papers gave authorization to obtain offers for reindeer, and if necessary to procure the number of deer that the relief expedition might lack to make a total of five hundred. The judges noted that neither Max Jansen nor any representative from his brokerage firm had inquired from William Kjellmann (the expedition's buyer in Finnmark) as to the number he lacked of reindeer, but rather, Jansen had proceeded to order from Sivertsen as many as five hundred deer only on an assumption that Kjellmann would not be able to fulfill the War Department's needs. In the opinion of the judges, Jansen's order thus was to be considered as independent and in no manner involving the War Department. The courts acquitted the U.S. Department of War, which had assumed on 9 May 1898 the part of defendant in place of Lieutenant Devore, of all charges but ruled that the department must share in paying court costs. Mocatta and Jansen were judged fully liable for the claims of Rivertz and Sivertsen, and were ordered to pay the plaintiffs as demanded in their itemized statements of financial losses (24,506.35 kroner and 3293.20 kroner, respectively), plus interest of four per cent annually (beginning 10 February 1898), and court costs (of approximately 952 kroner, in payment of which the War Department would share). In addition, the five thousand kroner deposited in February 1898 by Jansen with the Christiania police department were forfeited to Rivertz. The total amount payable by Mocatta and Jansen was approximately nine thousand dollars (U.S.) (not including interest).[97]

Sheldon Jackson and Daniel Devore, as the leaders of the Yukon Relief Expedition in Europe, both felt, uneasily, in January 1898 that the brokerage agency of Mocatta and Jansen in London was inclined more to take advantage of them by independent action and disregard for transactions in progress than to work with them according to plans of the War Department being carried out under direction of Jackson; but to define their apprehensions was difficult, since the brokers' mode of operation was somewhat covert—i.e., they appeared to omit information. The brokerage agents themselves failed to understand the plans of the two Americans, and they went beyond a concern to assist when Devore, following the orders of his commander the secretary of war, asked them to locate some reindeer in case a sufficient number could not be secured in Finnmark, after he had received a telegram from Kjellmann reporting that no deer would be available there until mid-February. Why Kjellmann sent such information is not known, since at the time the message was transmitted, he was buying or had reserved deer; that it was sent contributed exceedingly to the difficulty of Devore's assignment with the expedition.[98] Congress had provided $200,000 to aid the people in the upper Yukon River area and had turned to the Department of War for leadership in a national effort to provide that assistance at the time thought to be that of their most urgent need—late winter of 1898. Devore was given sole responsibility for administering the amount to be spent in Europe. While Dr. Jackson went north on the continent to carry funds to William Kjellmann in Finnmark, Devore stayed in England to deal with a most important element for a successful expedition—to find quickly and certify as appropriate and seaworthy a steamship to carry the reindeer and personnel to America. Without any possibility to consult with Jackson (although Secretary Alger clearly had ordered that the two would confer on their actions), he accomplished that with the desired economy, in spite of aggressive brokers. Undeniably, Devore made some unfortunate decisions, one of which led to an embarrassing and potentially disastrous court case.[99] Another, probably of greater significance to the expedition, was his directive that the *Manitoban* sail from Trondheim without loading more of the reindeer moss. Nonetheless, Secretary Alger fully supported Lieutenant Devore, who had followed official instructions and had carried out his assignment with high concern for its purpose.

After hearing Devore's report on 20 February, Secretary Alger ordered him to continue on special duty with the relief expedition. Since the *Campania* had required only a little more than a week to cross the Atlantic, Devore easily was able to meet with the secretary of war and return to New York City before the *Manitoban* arrived. He had ample time to make sure that all was in readiness there for the ship, so that the official inspections might be accomplished expeditiously and that debarking and transferring reindeer and personnel to the waiting railway cars would be managed smoothly. Secretary Alger wanted him to be present when the ship docked, to assist Dr. Jackson and perhaps to assume responsibility so that Jackson would be able to report immediately to War Department headquarters in Washington.

The *Manitoban* was on her way, slowly but to be given a certain dignity after coming through a great storm at sea, battered but intact, although on the day (20 February) that Lieutenant Devore reported to Secretary Alger and later took up his duties in New York City, she was having to shut down engines for repair, somewhere out in the Atlantic Ocean, to the alarm of the passengers in steerage.

# NOTES

## Section I.

1) Jackson, 1898b. The work was published as *Senate Document No. 34, 55th Congress, 3d Session (Serial 3728)*; a less detailed account (Jackson's eighth annual report on the introduction of reindeer) appeared in 1899 in the Report of the Commissioner of Education, Chapter XLI, pages 1773–1796, in Vol. 2, *Annual Reports of the Department of the Interior for the Fiscal Year ended June 30, 1898.*

2) Funds from the Department of War supported some of the personnel at Eaton Station.
Concerning establishment of vocational schools in Alaska, Jackson (1893) had reference to the congressional acts of 2 July 1862 (to benefit agriculture and the mechanical arts), of 2 March 1887 (to provide agricultural experiment stations), and of 30 August 1890 (for support of agricultural schools) in his remarks (p. 1300): "These acts of Congress require the assent of the legislature of the State or Territory in order that their provisions may become available. But Alaska has no legislature, and is governed directly by Congress. On this account, and partly because nineteen-twentieths of the children to be benefitted belong to the native races, Congress has committed to the Secretary of the Interior the duty of making 'needful and proper provision for education in Alaska.' I would therefore recommend that an application be made to Congress to direct the Secretary of the Interior to extend to Alaska the benefits of the agricultural acts of 1887 and 1890, and secure the establishment of a school that can introduce reindeer into that region, and teach their management, care, and propagation, and also to conduct a series of experiments to determine the agricultural capabilities of the country."

3) In 1898, Sheldon Jackson was a member of the Board of Home Missions of the Presbyterian Church of the United States. At the time of his appointment to the federal service (10 April 1885), he was superintendent of Presbyterian Missions in the Territories. He had been concerned about the conditions of the indigenous people of southeastern Alaska, had found them to be unfairly treated by merchants from the United States (many of whom urged the people into an exchange of their furs, works of art, and other desirable goods, for alcohol), and in need of education to "learn the white man's ways," or "civilized" life (terms used by Jackson and many others). Under Jackson's leadership, the Presbyterian Church established the first United States' school in the Alaska District, on 10 August 1877, at Fort Wrangell; Mrs. A. R. McFarland was the teacher. (Jackson, 1879.) Several biographies of the life of Sheldon Jackson have been published; two, dealing respectively with his work in Alaska and his earlier career in the service of the church, are Lazell (1960) and Bender (1996).

4) In Unalakleet, the pastor at the church supported by the Swedish Evangelical Mission Covenant in America was Axel E. Karlson. He, his spouse, and two others, Malvina Johnson and Alice Omekejook, were the main teachers.

5) On 29 July 1898, the *Louise J. Kenney* arrived at Unalakleet with thirty-five (including eight children) of Jackson's company from Norway. Kjellmann was then superintendent of the Reindeer Project in western Alaska, Bureau of Education, Department of the Interior.

6) The mountains are the Nulato Hills (highest elevation approximately 3400 feet). Reindeer moss: lichens mainly of the genera *Cladonia* and *Cetraria*. According to John Muir (in his journal written in 1881, published in 1917), Chukchi herders knew about coastal areas near St. Michael where rich growth of lichens would provide favorable conditions for reindeer. Kjellmann, when he was supervisor at Teller Reindeer Station, probably also learned from the local people about such areas and was thereby influenced in his selection of the site for Eaton Station.

7) "Unalakleet" (in the southern Malimiut dialect of the Iñupiaq language: Ungalaqliit) has the meaning of "south wind blowing." (Leonard Brown, personal communication; also see Jacobson, 1984.)

## Section II.

8) Jackson as well as Dr. N. H. R. Dawson (United States Commissioner of Education) listed and discussed the schools present in the District of Alaska in 1888 (Dawson, 1889; Jackson, 1889). Also see Hulley, 1958. On the Pribilof Islands, the Alaska Commercial Company was obligated to establish and maintain schools as part of the agreement with the Department of the Treasury, awarding the company rights to hunt and sell pelts of the fur seals *(Callorhinus ursinus)* there.

9) Congress did not release the educational funds until August 1886. That year, on the chartered schooner *Leo*, Jackson travelled as far as Unga Island to fulfill his duties of inspection as general agent of education in Alaska (Jackson, 1887).

10) Captain Healy was born in Georgia, his mother a slave, the property of Michael Morris Healy, his father. Some years after the cruise of summer 1890, the captain was investigated for harshness in punishing intransigent crewmen (officers) aboard his ship and was denounced (charged with drinking alcohol while on duty) by the Women's Christian Temperance Union and the Coast Seaman's Union; he was fully exonerated by a board of review (Cocke and Cocke, 1983; Williams, 1987). For his important contributions during more than thirty-five years of duty as commanding officer on ships of the Revenue Cutter Service in Alaska waters, Captain Healy received no formal recognition until almost a century after his death (30 August 1904); in November 1997 the United States Coast Guard launched the icebreaker *Healy*, which was named in his honor. (U.S. Department of Transportation, Office of the Assistant Secretary for Public Affairs, Washington, D.C.: *News*, CG 34–97, 11–13–97.)

11) See pages 1261–1296 in Jackson, 1893; pages 1705–1731 in Jackson, 1895; pages 1091–1128 in Jackson, 1906. Jackson was lifelong friend of Captain Healy and his wife, Mary Healy.

12) Jackson, 1893, 1906. Date given by Jackson for his 1890 cruise was incorrect.

13) Bockstoce, 1986; see pages 26, 131, 335.

14) Jackson, 1893.

15) The first purchase of reindeer in 1891 was made through barter of goods; see pages 952–957 in Jackson, 1894.

16) See Jackson, 1898; Updegraff, 1908. In Congress, especially strong support for the reindeer husbandry program came from Senator Henry M. Teller of Colorado, in honor of whom Jackson named one of the first reindeer stations (that at Port Clarence, now the town of Teller) in

Alaska. In 1898, when war with Spain was imminent, the senator became known for his proposal (the Teller Resolution) of guarantee of independence for Cuba that attached to the declaration of war. During 1881–1885, Teller was secretary of the interior, in President Arthur's cabinet.

## Section III.

17) Jackson arrived at St. Michael on 27 June 1897; next day, the river steamer *Portus B. Weare* came in from Dawson City with a cargo of gold, which was transferred to the steamer *Portland* and carried to Seattle; there, news of the great fortune in gold brought an effusion of excitement amongst the people of Washington State and elsewhere. (See Jackson, 1898.) For Jackson's survey for the Secretary of Agriculture, see pages 133–140 in Jackson, 1898b.

18) See page 1642, Jackson, 1898.

19) The U.S. Army established a post at St. Michael, where on 8 October 1897 two officers and twenty-five enlisted men arrived, under command of Lt. Col. George M. Randall, Eighth Infantry. The Department of War also sent Capt. Patrick Henry Ray and 1st Lt. Wilds P. Richardson to the area of the upper Yukon River in August 1897. (See U.S. Dept. of War, Annual Reports, 1898.)

20) *Victoria Daily Colonist*, 27 October 1897, page 5. Reverend Gambell's news about the whaling fleet was the first credible report to be received in Canada and the U.S. Also see Jackson, 1899.

21) A refuge station built at Cape Smythe (Barrow) in 1889 for use of stranded seafarers had been closed in 1896. (Cf. U.S. House of Representatives, 1889; Bockstoce, 1979.)

22) *San Francisco Chronicle*, 27 October 1897, page l; 29 October 1897, page 2; 4 November 1897, page 5, et seq.

23) President McKinley met on 8 November 1897 with Secretary of War Russell A. Alger, Secretary of the Treasury Lyman J. Gage, Secretary of the Navy John D. Long, and military and naval men. (Report in the *New York Times*, 9 November 1897, page 5. Also see page 22, Jackson, 1898b.)

24) U.S. Department of the Treasury, 1899; Brower, 1960. Lieutenants Jarvis and Bertholf and Surgeon Call received congressional medals of commendation for their courageous efforts to assist the whaling crews; William T. Lopp, Charlie Antisarlook, and their assistants on the trail were recognized through monetary award. Charles D. Brower was a resident at Barrow and knew the people and area; his descriptions of the wreckage of the whaling ships and arrival of the reindeer as relief food supply contribute to the understanding of those events. (In 1915, holding the rank of commodore, Bertholf became the first commandant of the Coast Guard.)

## Section IV.

25) The spelling "Klondike" was derived from the Athabascan (Han) name for the river Thron-duick ("hammer-water"), which referred to the stakes that were driven into the river's bed to guide salmon into fish traps (Rayburn, 1994). The Yukon District was made a Territory

of the Confederation by act of Parliament, 13 June 1898. Estimates vary concerning the number of people in the area of Dawson City in the winter of 1897–1898. Captain P. H. Ray stated that at least six thousand would be destitute there that winter (Ray, 1899). Sheldon Jackson judged that five thousand were in need in December 1897 (Jackson, 1897a). For the entire upper Yukon River region at that time, a population of near eight thousand could be a reasonable estimate.

26) Captain Ray and Lieutenant Richardson had not been ordered specifically to visit the Klondike. They managed to travel as far as Circle City, beyond which about 300 miles on the Yukon River (then in shallow channels amongst treacherous sandbars, and freezing over) remained, to reach Dawson City. During the autumn they received no communications from the war department; mail sent from Seattle in September arrived in Circle City the last of December, October's mail came even later, and they did not know that Secretary Alger hoped to have information in October about what food was available in Dawson City. They established winter headquarters at Fort Yukon, from where their numerous outgoing communications were delayed as much as those incoming. Ray's dispatches in October through December went overland (with carefully chosen messengers) to Ely Weare and John Healy of the North American Transportation and Trading Company in Dawson City. Weare and Healy held them until someone, a courier they could trust, would carry them (on foot) via Chilkoot Pass to Skagway; the letters then went by ship to General Merriam at Vancouver Barracks, Washington, and finally on to Washington, D.C. On 5 January 1898 Ray sent word to Secretary Alger that at Fort Yukon and Dawson the people had sufficient food. His dispatch probably did not reach Alger until the end of March.

Evidently, not one of Ray's communications sent during autumn 1897 was received at headquarters of the department before some time in late February or early March 1898. If his reports had arrived opportunely, they would have lent strong support to the plan for a mission of relief, as they described in detail his and Lieutenant Richardson's efforts to maintain civil order as destitute miners moved out of the Klondike to seek food at the stores of the North American Transportation and Trading Company and the Alaska Commercial Company in Circle City and Fort Yukon (which was nearly 400 miles northwest of Dawson City). Robbery and theft, with and without use of firearms, were commonplace. When armed men in groups of 35 to 100 threatened to remove food by force from the stores, Captain Ray in the name of the United States government took possession of all supplies, including food, of both companies.

Not all of the needy were lawless. Many lacking everything but good intentions were put to work cutting wood under a program set up by the two officers, for which labor the men were to receive $5 per cord in food or money. Those not able to work received food at no cost but with the requirement that they be responsible for their own shelter. Men wishing to prospect locally were given provisions, on condition that each sign a note of indebtedness. On 21 January 1898, Ray stated that "The tide of lawlessness is rising all along the river," but that "The policy of forcing the people to cut wood at this point is having the best results, as in a distance of 50 miles there is now cut and banked over 5,000 cords, which will be increased to 10,000 by the time the river opens. This will greatly facilitate getting supplies up the river, as I learn it is the only point between here and St. Michaels [*sic*] where any amount sufficient to meet the demands has been cut." Wood was the fuel of all river steamers on the Yukon in 1897–1898. Captain Ray's effective actions to control the miners were taken independently and were not known to the Department of War until late spring 1898. To the adjutant general (1 November 1897), he expressed concern that he was proceeding without orders: "I only hope the President and Congress will sustain my action and treat me with charity should I be found in error." Neither he nor Lieutenant Richardson was judged to be at fault. The financial obligations incurred were reimbursed to the Alaska Commercial Company out of the congressional grant for aid to the miners (Ray, 1899a).

27) People had genuine concern for the gold miners. Many wrote to the War Department, giving free advice about how to proceed; some of course hoped for reward. The letters are all on file in the National Archives (War Department, AGO File 67221). As pack animals, burros were suggested, and Shetland ponies, and angora goats that would come with their own herding dogs from Forest Grove, Oregon. The top chef at the Palatine in Newburgh, New York, proposed that oyster meal would excel as food on the trail as well as in Dawson; he implied that the Palatine would supply this nutritious product, made from fresh oysters, eggs, and crackers, and dried, so that one had only to add water to make delicious cakes. The department did not award contracts for these.

28) *The Oregonian*, Portland, Oregon, 28 November 1897.

29) United States Congress. Senate, 1897a. Secretary of War Alger's communication to the Senate read in part: "The only possible routes by which supplies can be transported into the mining district at the present season would be either by the Chilkoot or White Pass...or through the Chilkat Pass and over Dalton's Trail.... From the best information that can be obtained it is believed that the use of reindeer will be the means by which these supplies can be gotten through, if at all. It is therefore recommended that reindeer be purchased in Lapland to the number of 500 and permission granted to bring reindeer drivers from that country.... It is believed that supplies taken into that country need not, to any great extent, be furnished as a gratuity, but that many of the miners will be able to pay the cost of such supplies." (Alger, 1897.)

30) Jackson, 1897a. Jackson's reports were convincing. See also Cannon, 1897; as chairman of the Committee on Appropriations, Mr. Cannon was a powerful influence in the House of Representatives.

31) Bailey, 1897. Mr. Sayers became governor of Texas (1899–1903); in the House in December 1897, some of his colleagues already were addressing him by the title of that office.

32) The question of food for the reindeer would eventually be a great problem, but the deer upon arrival in Seattle were in excellent condition (Jackson, 1898b).

33) United States Congress. House of Representatives, 16 December 1897.

34) Representative Sayer's words were significant in view of the trouble that U.S. citizens were having in Cuba. War with Spain was not far in the future.

35) H.R. 5173 as transmitted by the Department of War read as follows:

> Headquarters of the Army, Adjutant General's Office, Washington, December 28, 1897. General Orders No. 74. The following act of Congress is published for the information and government of all concerned:
>
> An Act Authorizing the Secretary of War, in his discretion to purchase subsistence stores, supplies, and materials for the relief of people who are in the Yukon River country, to provide means for their transportation and distribution, and making an appropriation therefor.
>
> *Be it enacted by the Senate and House of Representatives of the United States of America in Congress assembled,* That the sum of two hundred thousand dollars is hereby appropriated out of any money in the Treasury not otherwise appropriated, to be

expended (or so much thereof as may be necessary) in the discretion and under the direction of the Secretary of War for the purchase of subsistence stores, supplies, and materials for the relief of people who are in the Yukon River country, or other mining regions of Alaska, and to purchase transportation and provide means for the distribution of such stores and supplies: *Provided*, That with the consent of the Canadian Government first obtained, the Secretary of War may cause the relief herein provided for to be extended into Canadian territory.

That said subsistence stores, supplies, and materials may be sold in said country at such prices as shall be fixed by the Secretary of War, or donated, where he finds people in need and unable to pay for the same.

That the Secretary of War is authorized to use the Army of the United States in carrying into effect the provisions of this Act, and may, in his discretion, purchase and import reindeer and employ and bring into the country reindeer drivers or herders not citizens of the United States, or provide such other means of transportation as he may deem practicable. The said reindeer or other outfit may be sold and disposed of by the Secretary of War when he shall have no further use for them under the provisions of this Act, or he may turn over the same or any part thereof to the Department of the Interior, and the proceeds arising from all sales herein authorized shall be covered into the Treasury.

Sec. 2. The Secretary of War shall make report in detail to Congress at the beginning of its next regular session as to all purchases, employments, sales, and donations or transfers made under the provisions of this Act.

Approved, December 18, 1897.

By command of Major-General Miles: H. C. Corbin, Acting Adjutant-General.

(War Department, AGO File 67221, National Archives.)

36) Edward Hegner, a businessman in Seattle, upon return from a successful season of gold mining in the Klondike area, judged that ten million dollars in gold would be produced by the mines in 1897–98 (winter and spring) in spite of a food shortage (*Seattle Post-Intelligencer*, 28 October 1897, page 4). By 1900, estimated dollar amount (U.S.) of gold from the Klondike area was twenty-two million, most of which came into the United States.

37) United States Congress. House of Representatives, 16 December 1897.

## Section V.

38) *Saami*: The term (derived from Saami language) now preferred as the name, in English, for the indigenous people of northern Fennoscandia and adjacent Russia, formerly called "Lapps." The term need not be grammatically inflected. (See remarks by Myrdene Anderson, Consulting Editor, in Vorren, Ø., 1994.)

39) Cf. page 109, Jackson, 1898b. Also, see letter (December 1897) from Jackson to the Secretary of War (War Department, AGO File 67221, National Archives).

40) In advance of the congressional debates on the question, Kjellmann had travelled to Norway as escort for a group of Saami who were returning to Lapland after working in Alaska for three years (1894–1897) as instructors in the program of reindeer husbandry of the Bureau of Education. Undoubtedly, Secretary of War Alger knew that Kjellmann was already in Europe;

the secretary had called him for consultation in November 1897 in Washington, immediately before he left for Finnmark. (See report in the *New York Times*, 1 December 1897, page l; also page 32 in Jackson, 1898b.)

41) The request that Jackson be detailed for temporary duty with the War Department was made by the acting secretary of war (G. D. Meiklejohn) (War Department, AGO File 67221, National Archives). Jackson's orders from Secretary of the Interior Bliss were given in Jackson, 1898b (page 103).

42) The orders of the secretary of war to Jackson are on file in the National Archives, certified by the Adjutant General's Office, 5 January 1898 (War Department, AGO File 67221).

On 3 August 1897, Lieutenant Devore reported for duty in Washington, D.C., to serve as military aide to the secretary of war; in December, Secretary Alger assigned him to temporary duty with Jackson. Devore received a copy of Alger's letter to Jackson, and it is included in his official file at the National Archives (File of D. B. Devore, Military Reference Branch, National Archives). Upon that assignment, Devore received a letter of credit; from the congressional appropriation of $200,000, Treasury Warrant No. 1913 was issued on 24 December 1897, showing that Requisition No. 7477 drew out of United States funds $60,000 and credited that amount in his name ("Lieut. D. B. Devore, Special Disbg. Agent for War Dept.") to an account in Stockholm, sent in care of the U.S. Minister to Sweden. (U.S. Dept. of the Treasury, 1897.)

43) The Special Passport was No. 10435, which read: "United States of America, Department of State, To all to whom these presents shall come, Greeting: Know Ye, that the bearer hereof, Sheldon Jackson, a citizen of the United States, United States General Agent of Education in Alaska, is about to proceed abroad under instructions from the Secretary of War. These are therefore to request all whom it may concern to permit him to pass freely let or molestation, and to extend to him all such friendly aid and protection, as would be extended to like officers of Foreign Governments resorting to the United States. In testimony whereof, I, John Sherman, Secretary of State of the United States of America, have hereunto set my hand and caused the Seal of the Department of State to be affixed at Washington, this 22nd day of December, A.D. 1897, and of the Independence of the United States of America the 122nd." (Jackson, Travel Journals, 1897–1898, in the Archives of the Presbyterian Historical Society, Philadelphia.) Neither Jackson nor Devore had visited Europe prior to 1898.

44) Secretary of War Alger indicated that Jackson was to receive a stipend of one hundred dollars per month while he was on duty with the War Department; accordingly, he was paid for thirteen months of his service. He continued as Special Agent for an additional nineteen months. In Europe, amounts to cover expenses were allotted by Lieutenant Devore. (Jackson, Correspondence, 1897–1898, in the Archives of the Presbyterian Historical Society, Philadelphia.)

45) Theodore Ayrault Dodge, brevet lieutenant colonel, United States Army, retired, was a historian and author of numerous books of military history, including *The Campaign of Chancellorsville* and the famous series *Great Captains*, in which Dodge presented his studies of renowned military leaders and their plans and tactics.

46) The Saami had a sound policy concerning money. In a letter to Secretary of War Alger (7 March 1898), Jackson commented on their attitude: "They have no confidence in a promise to pay. A would-be buyer must show his money or at least make partial payment...." The letter is on file at the National Archives and published in Alger, 1899.

47) Jackson, Travel Journals, 1897–1898. Presbyterian Historical Society, Philadelphia.

48) On Sunday (2 January 1898), the one day in London holding no obligation for duty, Jackson hoped to be at St. John's Presbyterian Church when some Korean converts to Christianity were to be baptised, and at Westminster Abbey to hear the sermon of Rev. Canon Wilberforce. He took the bus (to save money), and it crept along so sluggishly that the ceremony was over when he arrived; but his friend Rev. Gibson acknowledged the presence of a colleague from America and brought Rev. Dr. Jackson forward to assist in the Eucharist and to say a few words to the congregation. The bus was slow again on the way to Westminster Abbey, and he missed the sermon. In the evening, he and Lieutenant Devore walked to the worship service at the London City Temple. (Jackson, Travel Journals, 1897–1898. Presbyterian Historical Society, Philadelphia.)

## Section VI.

49) Jackson recorded (page 33 in Jackson, 1898b) that the disease was "foot-rot"—a rather common condition in reindeer (see page 6, Qvigstad, 1941). Probably, however, the cause for the quarantine was a viral infection, foot-and-mouth disease. In his biography of Jackson, Lazell (1960) stated (page 162) that the condition was "hoof and mouth disease" and that Great Britain excluded by quarantine even the ships that had carried livestock from Scandinavia to England. Outbreaks of foot-and-mouth disease in domestic ungulates may cause severe economic loss. The infection in deer usually comes through transmission from cattle (see Nikolaevskii, 1968).

50) Kjellmann had agreed to pay for five hundred loads of reindeer moss at seven kroner per load and four hundred loads at eight kroner (less than three dollars each). The brokerage agency of E. A. Tonseth had been recommended by Mocatta, Jansen and Company (see page 35, Jackson, 1898b). Tonseth's eventual brokerage fee was $8468.65 (U.S.), which included payment for reindeer moss, its shipment by railway from Røros to Trondheim, loading and unloading the railway cars, and salt and bedding (hay) for the deer. (Alger, 1899, page 41.)

51) Perhaps comparing conditions in Alaska with those observed on the journey from Trondheim to Tromsø, Sheldon Jackson commented, "A peculiarity of the Norwegian Coast is the great number of lighthouses. The steamer no sooner loses sight of one passed than another looms up ahead. Consequently, these narrow and dangerous channels can be navigated the darkest nights, the only trouble to the navigator being a thick blinding snow storm. Another peculiarity is the extent of postal and telegraphic facilities. Every important village up to North Cape has mail facilities and also either telegraphic or telephonic communication with the world. There are in Norway telegraph stations north of the Arctic Circle. Small steamers carry the mail to all the little villages along the fjords and the swift reindeer carry it inland hundreds of miles. But few of the places where the steamer carried the mail had wharves, but a large row boat would come out to meet the steamer, exchanging mail, freight, and passengers." (Jackson, Travel Journals. 1897–98. Presbyterian Historical Society, Philadelphia.) (The first lighthouse in Alaska was built in 1902.)

## Section VII.

52) Graarud and Irgens, 1921.

53) "Bossekop" (language of the Saami) may be expressed as "Whale Bay." (Øystein Boland, personal communication; also see Lagercrantz, 1939; and Jackson, Travel Journals, 1897–1898.

Presbyterian Historical Society, Philadelphia.) Alta is the site of a historic center of trade for the people of Finnmark—the Saami, Norwegians, and Finns. On the headland of Komsa near Bossekop, rock carvings and artifacts have been dated to ca. 5500 B.P. (Hood, 1988).

54) Jackson, 1898b. Also see Jackson, Travel Journals, 1897–1898. Presbyterian Historical Society, Philadelphia.

55) Devore graduated twenty-ninth in his class at the United States Military Academy (West Point). He was a member of the faculty at the Military Academy (assistant professor, Department of Mathematics) from 20 August 1892 to 31 July 1897, when he was appointed military aide to Secretary of War R. A. Alger. In addition to mathematics, he had special interests in literature, languages, and law; he was admitted (24 April 1890) in Detroit, Michigan, to the practice of law. Immediately after the Spanish-American War (in which he served), he was governor of the District of Lake Lanao and was cited by Major General Leonard Wood for his leadership and contributions to civil government in the Philippine Islands. During World War I, he was in command (brigadier general) of the 167th Infantry Brigade in France, 1917–1918. In 1915, Devore (then lieutenant colonel) on duty in Panama was asked why he had made no reports, and again he informed the adjutant general of the army (who had contacted him about the deficiency) that he had not been ordered to report. Devore was retired in 1930 at his highest rank (brigadier general during service with the National Army in France). (Files of Daniel B. Devore, Military Reference Branch, National Archives; also see Cullum, 1891 et sqq.)

56) The *Manitoban* originally was named the *Ottawa*, registered in London (No. 231, 1865). She began as a three-deck, three-masted schooner, iron framed, built in 1865 at Birkenhead for the British Colonial Steamship Company, London-Quebec-Montreal. In 1872 (after purchase by the Allan Line of Glasgow and Montreal), she was altered to four decks, two masts, one funnel, lengthened from 237 feet to 339 feet and compounded, and renamed the *Manitoban*. Her official number was 52746, registered in Glasgow, 1887. Gross tonnage 2974 tons. She was used mainly on service from Glasgow to the United States. The voyage of 1898 with the reindeer was near her last; on her certificate of British registry is written "Sold to Italian interests. Cancelled and Registry Closed. 5th June 1899." Shortly after, she was broken up for salvage. (Information from the files of the Strathclyde Regional Archives, Glasgow, and of the Public Record Office, Richmond, Surrey.) Also see Appleton, 1974. According to J. M. Wraight, Archivist at the National Maritime Museum, London, the log book records for the Allan Line have not survived (personal communication).

Lieutenant Devore rightly judged the *Manitoban* seaworthy; nonetheless he did not pay at once for the round-trip charter, but he waited to see that she arrived in New York harbor before he gave Allan Line the second half of the fee. Total cost of her charter was $9100.00 (U.S.). The Allan Line received an additional $6359.17 (U.S.) for provision of food for the passengers and care for the reindeer (labor only to tend the deer; reindeer moss, obtained in Norway, had been purchased in advance). (Alger, 1899.)

57) See CJS Journal.

58) Board of Trade, 1898. (Ship's manifests were provided by the Maritime History Archive, Memorial University of Newfoundland at St. John's.)

59) For harborage records, see *Lloyd's Weekly Shipping Index,* 1898. Information concerning the ship's stop at Bergen was provided by Ms. Kari Helgesen, Statsarkivet i Trondheim. Lieutenant Devore went by train on 11 January to Newcastle-on-Tyne, and from there by steamship to Trondheim via Bergen. (Files of Lt. Devore, Military Reference Branch, National Archives.)

He made no report for the week in Trondheim prior to arrival of the *Manitoban*, but Jackson requested that during that time, he arrange for loading the reindeer moss on the ship.

60) Secretary of War Alger was ill and unable to be at his office during much of January and February. Assistant Secretary G. D. Meiklejohn served as acting secretary of war at that time.

61) In his personal logbook, Dr. Jackson wrote in a notation dated 3 January 1898 that an agreement was made with Mocatta and Jansen providing that the brokerage agents were to give assistance in arranging for a steamship and procuring deer, in return for which they would be paid three hundred dollars, or if it became necessary for one of them to go to Norway, five hundred dollars; they were to pay their own expenses for such travel. That notation may have referred to the following letter addressed to Jackson and Devore, signed by Mocatta, Janson and Co.:

> London. 3 Janary 1898.
> Gentlemen. Confirming the arrangement we came to this morning, we have pleasure in herewith handing you a short agreement for yr signature. As we now understand the case it is settled on the following terms. That we are to consider ourselves under direct orders from you and that only on such orders are we to act. We have intimated to our representative in Mosjøen that he should purchase 300 reindeer, we also instruct this agent and another one in Drontheim to purchase a quantity of moss. We should like this confirmed by you, whether these orders are to be carried out, it having struck us since our conversation this morning that you wish to have a certain quantity of reindeer at your option of purchase at Mosjøen. This option cannot be expected to be extended to you, in as much as our friends will probably have the whole trouble for nothing. Therefore it would be well if a decision is come to as we could then wire Mosjøen to cancel the order. To us it is very apparent that to further an early shipment, the more that is done from this side the better, but will leave everything entirely in yr hands. Reports that we have from Norway seem to be so contrary to what Mr. Kjellmann have advised you. Why our friends should say Mr. Kjellmann is not getting many reindeer we don't understand, still we think it better to mention this to you.
> Yours faithfully....

No response to the letter is indicated in Jackson's file. (Jackson, Travel Journals, 1897–1898. Presbyterian Historical Society, Philadelphia.) The agreement was dated 4 January 1898 and signed by Devore.

62) Letter from Lieutenant Devore to Secretary of War Alger, 10 January 1898; in File of Lt. Devore, Military Reference Branch, National Archives.

63) Communication from Lieutenant Devore to the Office of the Adjutant General, 18 March 1898, in File of Lt. Devore, Military Reference Branch, National Archives. Documents in the archives (personal documents of Dr. Jackson) at the Presbyterian Historical Society, Philadelphia, include the following telegram from Max Jansen to Jackson:

> Trondhjem 13 Jan 1898 2.35 PM. Dr. Jackson Alten. Kjellmann made most abominable mess. Whole business discredit country round. Telegraph me how many reindeer, moss, Lapps you have ready shipment 22d instant. I have bought 500 reindeer, 250 Tons moss for that dau [*sic*] for whom has Kjellmann been shipping by Thingvalla 150 reindeer. You must ere this see I know my business and see all I said

in London is borne out by facts. Keep nothing back if you want to further the interest of the United States government. Janson Grand.

On the telegram Dr. Jackson wrote "Made no reply to above. SJ." Perhaps the communication was Max Jansen's fulfillment of Lieutenant Devore's instructions that Jackson be informed of any negotiations concerning reindeer and fodder. The telegram was not published with others (see Jackson, 1898b).

64) The buyers sent out by Kjellmann were Mathis Eira, Samuel Kemi, M. Kjeldsberg, A. Paulsen, Per Rist, and Carl Suhr. Paulsen was a merchant at Karasjok in Finnmark; Eira, Kemi and Rist had worked in Alaska for the Bureau of Education.

65) Copies of telegrams are in the records of the War Department, AGO File 67221, National Archives. Also see pages 109–119 in Jackson, 1898b. Among the telegrams preserved is that of 11 January 1898 from Lieutenant Devore addressed to Jackson in Hammerfest; evidently, no record exists of a response from Jackson.

66) Reported in the newspapers *Dagsposten*, Trondheim, 22 January 1898, and *Trondhjem's Adresseavis*, 24 January 1898. (Personal communication from Ms. Kari Helgesen, Statsarkivet i Trondheim.)

67) Telegram from Devore to Jackson, 24 January 1898: page 118 in Jackson, 1898b.

68) *Trondhjem's Adresseavis*, 5 February 1898.

## Section VIII.

69) Jackson did not understand how people existed with some comfort in arctic regions; his experience in northern Alaska had been during months of summer only (as had that of most other officials of the U.S. government at the time). His introduction to winter in the far north took place at Alta. He went to the church, a convenient place to view the town's residents. He noted that beneath the hem of the splendid liturgical robes of the priest protruded his fur shoes, wide, ungainly and incongruous, with toe part bent upward in a hook (they were made in Finnmark, of Saami-style, ideal for cold weather), and that from time to time during the service, the precentor spit tobacco juice over the railing of the choir onto the church floor. Comprehending neither the sermon nor the songs, Jackson wrote somewhat doubtfully that he hoped that "the people were spiritually fed." (Jackson, Travel Journals, 1897–1898. Presbyterian Historical Society, Philadelphia.)

70) Jackson, 1898b.

71) Note of 29 January 1898: Jackson, Travel Journals, 1897–1898. Presbyterian Historical Society, Philadelphia.

72) See pages 38–39, in Jackson, 1898b. Antlers had been removed from the deer before they were placed on the ship. The reindeer were males, all specially accustomed to harness and sleigh. Kjellmann purchased sleighs at $3.50 each and harness outfits at $3.00 each; the cost of each reindeer was approximately $10. (Jackson, 1898b, pages 39 and 145.)

73) The contract in full was given in Jackson, 1898b (pages 107–108), and in Alger, 1899 (pages 51–52). The abbreviated format was signed by fifteen men. (War Department, AGO File 67221, National Archives.)

74) In the records at the National Archives (AGO File 67221), the contracts were found to be tied and fastened together in one unit; that of Carl Johan was torn and partly missing.

75) Personal communication from Arne G. Edvardsen, Statsarkivet i Tromsø. In the church register of Karasjok Parish, Mr. Edvardsen found no records of emigration to Alaska in 1898.

76) Beach, 1986.

77) Jackson, 1898b.

78) *Finnmarksposten* (Hammerfest), 15 February 1898. (Information from Ms. Ida Bull, Statsarkivet i Trondheim.) In the newspaper's account, some errors occurred in spelling of names and origin (home) of contractees.

79) Those returning to Norway in December 1897 were some of the people hired earlier (1894) to work as instructors in the reindeer husbandry program of the Bureau of Education. In 1894, prior to his arrival in Alta, Kjellmann had contracted with the Feddersen and Nessen Agency in Hammerfest to advertise in the regional newspaper(s) for experienced reindeer managers who would take temporary employment in Alaska. Jackson could not accompany his assistant to Norway that year, but he was pleased with the personnel Kjellmann had chosen. (Jackson, 1896.)

80) When the Saami asked for opportunity to make inventories of their personal possessions before departure, it was said to be a tactic for further delay; but possibly the request was misjudged, inasmuch as such care before travelling may have been usual amongst the Saami.

81) Jackson, Travel Journals, 1897–1898. Presbyterian Historical Society, Philadelphia.

82) See CJS Journal; also page 39 in Jackson, 1898b.

## Section IX.

83) Jackson, 1898b; also see Jackson's notes: Travel Journals, 1897–1898. Presbyterian Historical Society, Philadelphia.

84) Devore's telegram(s) to Jackson concerning the legal action were not found in the archives. Jackson's telegram to Devore did not indicate the matter(s) about which the American consul was to be informed.

85) In addition to the published eight telegrams (in Jackson, 1898b) from or to the brokerage agencies, others are in the records of the War Department, AGO File 67221, National Archives, Washington, D.C., and in the archives of the Presbyterian Historical Society, Philadelphia (Documents of Sheldon Jackson).

86) The information about the reindeer in Skibotn and their escape from the enclosure built specially to hold them was given in the documents of the Municipal Court in Christiania: Christiania byrett, domsprotokoller I 53, 1898–1900, s. 492–505, Arkivverket, Statsarkivet in Oslo. The deer may have been recaptured; about that, the language of the documents is not clear.

The *New York Sun* (see Annotations for 28 February 1898) reported that five hundred reindeer were aboard the *Thingvalla* when she arrived in New York (on the same day as the *Manitoban*), and that those animals had been purchased in Scandinavia by Max Jansen and shipped to the United States for speculation on the livestock market. According to *The World* (New York) (28 February 1898, p. 9), the *Thingvalla* brought about sixty deer, "in competition with the Government animals."

The disposal (and sale) in the United States of those animals could not be traced; but the possibility seems to exist that they were some of the same deer for purchase of which Rivertz and Sivertsen later were reimbursed, also, under the demands of the lawsuit brought by them, by decision of the courts of Norway in 1900.

87) Letters from Emanuel, Round and Nathan, and letter of 11 February 1898 from Max Jansen, to Secretary of War: War Department, AGO File 67221, National Archives.

88) Court Records: Christiania byretten, Sivile Saker, I–32, 1898; Ekstraretts protokoll nr. 34, s. 242–248; Tingbok 91, s. 79, 262. Year 1899: case 212–98, 213–98—s. 45, 79, 88, 126, 143, 175, 194, 229, 246–47, 284. Arkivverket, Statsarkivet i Oslo.

Concerning certain laws of Norway, see Andenaes, 1965; Evensen, 1965; Schjoldager and Backer, 1961.

89) Court Records: Christiania byretten, Sivile Saker, I–32, 1898, Ekstraretts protokoll nr. 34, s. 242–248, Arkivverket, Statsarkivet i Oslo.

90) Communication (2d Endorsement) to the Adjutant General, Department of War, from Lieutenant Devore, 18 March 1898: File of Lt. Devore, Military Reference Branch, National Archives.

91) Court Records: Christiania byrett, domsprotokoller I 53, 1898–1900, s. 492–505, Arkivverket, Statsarkivet i Oslo.

92) Letter of American Consul to Assistant Secretary of State, and Cablegram: File of Lt. Devore, Military Reference Branch, National Archives.

93) War Department, AGO File 67221, National Archives. Max Jansen's name at times was spelled Janson, Johnsen, or Johnson.

94) Letters of Consul Bordewich and Attorney Aubert, 25 April 1898, and from the Assistant Secretary of War to Attorney Schiander, 9 May 1898: File of Lt. Devore, Military Reference Branch, National Archives.

Ships' manifests for the *Manitoban* in spring 1898 indicate an exchange rate of one pound (Britain) to 4.86 dollars (U.S.A.). Reference: Board of Trade, 1898.

95) Court Records: Christiania byrett, domsprotokoller I 53, 1898–1900, s. 492–505. Arkivverket, Statsarkivet i Oslo.

Letters of Consul to Department of State: File of Lt. Devore, Military Reference Branch, National Archives.

Letters from Max Jansen: War Department, AGO File 67221, National Archives.

Letter from Ambassador Pauncefote to the Secretary of State: War Department, AGO File 67221, National Archives.

96) Communication dated 18 April 1899 from Assistant Secretary of War: War Department, AGO File 67221, National Archives.

The Mocatta and Jansen Brokerage Company received 2455.00 ($ U.S.) from the War Department. (See Alger, 1899.)

97) Court Records: Christiania byrett, domsprotokoller I 53, 1898–1900, s. 492–505. Arkivverket, Statsarkivet i Oslo.

98) See Jackson, 1898b. Lieutenant Devore later reconfirmed with Jackson that Kjellmann had actually sent the telegram.

99) A reporter from *The World* (New York) (28 February, p. 9) talked with Lieutenant Devore shortly after the *Manitoban* arrived: "Lieut. Devore was especially gratified at the safe arrival of the deer, as great difficulty was met with in procuring the animals owing to the alleged plot of one Hansen, [*sic*; Hansen = Max Jansen] who tried to corner all the reindeer in Lapland and sell them at a profit to the Government agents. Hansen was finally locked up in jail at Christiania, the charge against him, according to Lieut. Devore, being that of obtaining property under pretense that he was an agent of the United States." Evidently, Devore did not inform the reporter that he also had been under arrest in Christiania.

*Where he was, was home...*
—Rolf Baldwin,
in reference to Carl Johan, his grandfather

# The Journal of Carl Johan

On the early morning of 4 February 1898 when the *Manitoban* slowly turned in Altenfjord and headed out, away from Bossekop and Alta, not many of the passengers were asleep or even at rest. In the dim and ill-ventilated steerage area on the third deck, as in an auditorium, the all-night conversations were for anyone to hear, or to join, especially in wishing aloud, as many did, to go back home again. But the ship glided along through the dark hours, without a pause or hesitation, and to return was no longer possible. Carl Johan went up to the main deck, to keep vigil as the ship moved from point to point along the coast of Norway, and then to wave a last farewell to Fladstrand, his home on Langfjord. The waterways and shorelands he knew so well were almost an extension of home. He noted in his journal the route of the ship. The log books of the *Manitoban* no longer exist, but his record serves partially as one for the old vessel on her journey from Bossekop to New York Harbor in February 1898. The journal began on 2 February 1898.

# *Reisen til Alaska*
## Journey to Alaska—1898

Very often it is the case, when a young man leaves home on his first long journey, that it is a sad moment. He shakes hands with father and mother, brothers and sister, to bid them farewell, perhaps nevermore to see them in this life, and leaves friends and neighbors behind, to go out into the wide world in order to seek his own fame and fortune. If either of these is already determined for him, he begins the journey full of hope, but also uncertainty, to strive toward the goal; for it is always his plan that he will try with honor and toil to gain some means, then to have the joy of returning home to find his parents, siblings and friends still alive—to be united once again, and to enjoy the domestic pleasures as before.

When he leaves his home, it frequently happens that he begins a book known as a diary, in order to write down everything he experiences, every remarkable thing he sees, and every curious thing he hears. Out of that which will come later, there will never be a lack of material for the contents of a little book of recollections of the various places where one has traveled, since foreign travels to strange places are anything but monotonous.

Thus I decided, before my departure from home, to begin a brief diary or pocketbook, in order to write down every noteworthy thing that I perceived, that I would see and hear, especially since there were to be so many in the party. Everyone would want to recollect the Reindeer Expedition that the United States Government chartered from Alten, Norway, under the leadership of Dr. Sheldon Jackson and Mr. Wilhelm Kjellmann in the year 1898. They purchased 540 reindeer, but several were lost in transporting them aboard the ship, so that there were actually less than that number. In order to care for these animals during the voyage and in order to teach and instruct the Eskimos about the needs of reindeer, 65 men were hired, of whom 18 were Norwegians and 48 were Lapps from Lapland; 4 Norwegian-Finnish lads from Lakselv on Porsangerfjord came along, as well as one Lapp from the Trondheim area. Thus the entire company for the expedition, including women and children, numbered slightly more than one hundred people.

## February 2

— After I had been hired I left home on February second, 1898, planning to take a boat ride down the fjord. In the morning, however, there was such a strong wind blowing up the fjord that it was impossible to take the route by sea. Here was the first obstacle that would hinder my journey, but I still had one alternative: to hire some men to help carry my things across the mountains to Talvik, and possibly to arrive at Bossekop by this manner. I enlisted four men to go with me across the mountains. Although both storm and driving snow blew right against us, we attempted to brave it. But when we had traveled some distance in the mountains, one man became sick, and he froze both ears and his face and had to turn back, accompanied by another man. We three remaining continued to move along, constantly observing each other's faces and ears to see that they did not become frostbitten. In that manner we traveled through the storm and arrived at Talvik at 8 o'clock in the evening. We were tired and hungry, since nearly all of our food had gone with the two men who turned back from the mountains; we had competely forgotten about the food. We had covered forty kilometers.

## February 3

— I caught a boat from Talvik in the morning and arrived at Bossekop at midday, meeting Kjellmann just as the boat docked. He informed me that the ship was to depart that same evening. I had arrived at the last moment. I transferred all of my clothing aboard, stowed all of my gear, and was finished by 7 P.M. when I boarded, to make sure I should not be left behind. When I boarded ship in the area between decks, I became a bit discouraged at the appearance of sanitary conditions; but, undaunted, I sauntered to the room where all of the others were gathered. Our quarters were down on the third deck, which did not exactly appear to be luxurious. When the newcomers to the great Atlantic vessel *Manitoban* inspected her and saw how dark and black everything appeared, some refused to make the journey; they did not wish to travel aboard her. We would soon find out just what we had gotten ourselves into.

## February 4

— The night ended with the shouts and rough behaviour of some drunken Lapps. A number of people wanted to go ashore, but that was out of the question now. At 4 o'clock the *Manitoban* weighed anchor and sailed out of the Altafjord. I went to the main deck accompanied by my cousin, Wilhelm Basi. We waved our farewell to Langfjord, though no one noticed, while the ship slipped slowly out toward Stjernsund. It was not long before we experienced what conditions aboard ship would be like; we sensed that we were all a group of fellows without a will of our own. In the morning J. Tornensis became angry when he did not get food immediately, and he promised payment to all who would throw all of the Englishmen overboard and steer the ship back to Bossekop in order to get back ashore before it was too late. At dinner time we passed Loppen, where the pilot and customs officer were set ashore. From there the course was set for the open sea, and in the twilight the coastline sank below the horizon behind us.

## February 5

— In the morning, when we went to the main deck, we saw nothing but open sea on all sides. Although there was good weather and a light sea, many people were seasick. Unaccustomed to the sea, the Lapp women and children threw up in every corner, so that it was difficult to feel disposed to remain in the room with them. If one wanted to remain below decks, however, there was no place else to go, unless one chose to enter the dark moss storage rooms or to stand amongst the reindeer on the middle deck. During the day the reindeer were arranged in such manner that there were eight animals in each stall, and each man was in charge of the feed for one stall, or for eight reindeer. Later in the day, the steward offered to sell the Lapps a sort of soft drink. The Lapps were quite fond of it, paying 33 øre per bottle that was half the size of an ordinary Norwegian soft drink bottle.

## February 6

— Sunday. We had an easterly wind with heavy seas, and it was unpleasant on board, due to the incessant vomiting and the overpowering stench that prevailed in our compartment. This was not surprising since over 100 people were lodged together. After I had eaten my evening meal, I, too, became nauseated—seasickness was something I had never before been subject to. The food on board was nothing to boast about.

## February 7

— The Lapps are in poor condition due to seasickness. I prepared reindeer meat for those of my Lapp acquaintances who had any with them, for they all were sick; in return they permitted me to have some, too. It is difficult to eat the food that is prepared on board, since it is handled in such a foul manner by the cooks. Compelled by hunger, we sometimes have to eat it, although it is entirely disagreeable. We receive nothing else, regardless of what we do or say.

## February 8

— We awoke to the same existence as before, although the weather is calmer. When the sun eventually peeked through the clouds, I thought I felt a bit better than last evening. This evening we could see a navigation light to starboard, which indicated for us the southernmost tip of Iceland, and it disappeared in an instant. Now we knew that we were beginning to cross the Atlantic.

## February 9

— In the morning when we awoke, we saw to our joy that the ship was managing to falter along against the wind and waves. We knew that she did not travel faster than eight miles (per hour) in calm water, and that it would be impossible to make more than three and one-half miles (per hour) in a storm. The latter is the case, according to the crew. The sea steadily pounded the ship from stem to midships, so that it was not possible to go up on the main deck. In our quarters below, conditions were not the best. It was dark and dirty, unhealthy and wretched. In addition the sick people who could not move about at all were there. And when the ship pitched so that the stern rose from the water, the propellers spun dry with a sound that seemed at times to rend both the ship and our brains; but both held out through the day.

## February 10

— The storm continued to the west with doubled strength, increasing so that the ship made little progress. The Lapps passed the time by card-playing, which has been their entertainment during the entire journey.

## February 11

— No calm; the Atlantic promises us continued resistance, for the storm increased greatly until the spray was like fog. It was certainly most entertaining down in our stinking cabin. When these unseaworthy people try to move as the ship veered about, they try to grasp onto something stable, achieving some lovely acrobatic maneuvers in the process. At dinnertime a banging commenced to starboard, right beneath one of the lifeboats; one of the heavy iron support beams under the boat had broken. Out of curiosity, Ole Stensfjeld decided to go astern and observe how it looked when the propellers spin dry. And he was glad to come back alive, as he was nearly washed overboard to the sea. When he came below, he was wet from head to foot—the first to be baptized by the Atlantic. I was told that the ship covered one mile today, but I really do not believe it was that much.

## February 12

— The wind shifted more to the northwest, so that the sea was to starboard. After breakfast some of the women planned to go above in order to breathe a little fresh air. But just as they came out of the companionway, a great wave broke over the ship and poured over the entire group, and they had to return below decks at once, their reindeer tunics dripping from every fold. We had some small amusement from this.

## February 13

— Sunday. As far as I can recall, this is the most unpleasant Sunday that I have ever experienced. The storm raged, the ship heaved and groaned, and the food was unappealing, and unclean. In the evening one of the lifeboats was crushed to splinters by the sea, along with a number of planks from the reindeer stalls on the main deck. None of the animals was injured, however.

## February 14

— It is impossible to describe sanitary conditions aboard with regard to preparation of the food. The cook, a one-legged man, is a thorough-going pig, to my thinking. After he had chopped and cleaned the fish, I saw that he chopped meat on the same cutting board without first cleaning it; thus the meat becomes mixed with all the inedible fish entrails. And when he went to fetch meat from the ice cabinet, he climbed completely inside the cooler and trod squarely on the ox meat with his boots—the same boots that he wore when he went into the reindeer stalls. Accordingly, the meat took on another color. From what I observed, it looked as though the ox meat was thoroughly polluted by all the manure that the cook trampled into it each time he went to fetch more. The reindeer deck was filled with dirty water, urine and filth so deep that one had to wade up to one's knees in order to get through the enclosure. A good deal of this filth was spread to the meat in the previously mentioned manner. At dinnertime we raised a protest over the crude treatment of the food. Kjellmann was summoned, but Dr. Jackson, who sat astern in the ship's lounge, did not take one look at these conditions. At first we suspected nothing ill of our main boss, but we later thought that everything went according to his wishes only. No changes were made, although Wilhelm Kjellmann came down to our compartment; we had to make do with the food as it was set forth. With regard to sanitation, conditions down in our quarters were scarcely better than those in the kitchen. A Lapp woman bathed her child in one of the buckets used by the cook to carry soup for the passengers, and did this continually. When people became aware of this, many lost their appetites and tried to purchase something more edible from the ship's lounge, through the steward.

## February 15

— The storm has increased to such a degree that the Lapps have stopped playing cards. Many of us went about with grave expression, for this is certainly not amusing. I was a bit anxious, as were many others; for it often seemed as though the storm and the sea would rip the old *Manitoban* apart. Despite the ship's size, it began to feel as if it were quite light in the water; but to all good fortune our ship withstood the powerful hurricane. Only one lifeboat was crushed, along with some planks from the reindeer stalls on the main deck, but none of the animals was injured save one, which died after being hurt during the storm. The reindeer had to remain on deck all day without any food at all, or anything, save for the sea water that the storm drove over them incessantly, day and night. The captain said that this is the second time he has experienced such a hurricane, and he has been sailing on the Atlantic for twenty-five years. At night everything was closed and battened down. The night watch had stood at his post alone the entire night. The captain and his mate had gone below, leaving the ship to its own fate. We were told of this in the morning, by which time there was sunshine and good weather with no sign of distress or danger. The bow or figurehead was totally broken away, but the water did not spill into the lower decks due to a retainer that held out the sea.

## February 16

— Peace and sunshine have made us forget the perils of the storm that raged against us for over a week, hindering the ship from making the least bit of progress. The ship now continued its usual slow course westward. We spent the day strolling about the main deck and breathing fresh air, engaging in all sorts of amusements to make the time pass more quickly until we come to New York. We are all longing to go ashore, where everything is not so cramped and crowded.

## February 17

— For lack of other duties we were set to work at sewing up sacks in which to carry moss for the reindeer on the train journey across America.

## February 18

— When all of us had crept above deck to enjoy the fine weather, we saw to our astonishment Dr. Jackson's enjoyment as he came onto the main deck, took a little Lapp girl by the hand, and strolled forth and back across the area. We thought it a curious sight to see the chief walk stiffly about in his wolf-skin coat, holding the little girl's hand and smiling faintly.

## February 19

— There is an easterly breeze, so the ship's sails were set—but only two foresails, which did not help us in the least.

## February 20

— Sunday. Our dinner today consisted of some sort of soup, in which all manner of filth could be observed floating, including a piece of rotten, dirty ox meat and some potatoes, which had frozen on the main deck while the ship lay at Bossekop. We also received a type of roll, made by the ship's excellent baker. This was our fare at virtually every meal. As we were about to take a little after-dinner nap, the ship's engine stopped. I went above deck along with some of the others, to see what was the matter. Some said that the engine had to be stopped for cleaning and maintenance, since it had been in constant operation for two weeks. Many accepted this explanation, as none of us knew exactly what was wrong. During this confusion, one of the Lapps ran into the compartment and shouted that the engine had broken down. At this news, men and women sprang for the main deck with their children, nearly in tears from despair. To our unhappiness, the Lapp's report was verified—there had been damage to the engine. At this point it suddenly occurred to me just how dangerous it would be to be shipwrecked in the North Atlantic, away from all shipping lanes. But the damage was repaired in an hour's time, whereupon the old *Manitoban* once more began to creep along. The damage consisted of cracked insulation around one of the steam pipes.

## February 21

— All went as well as usual. Kjellmann gave orders that an inventory of personal goods should begin, and that it should be completed before our arrival in New York; thus we had a new activity to make the time pass more quickly, for it seemed always so long and monotonous.

## February 22

— It was still and foggy all day, so the ship's horn sounded every five or ten minutes. In the evening Kjellmann came down to our dark and lovely quarters and informed us that we had already crossed over into the Newfoundland Banks, where the sea is always ferocious during storms and bad weather, due to shallow water.

## February 23

— Everything was well on board, though many complained of the abundance of vermin. This was of little consequence, as we were all resigned to our fate. Nonetheless, one reflected that all things come to an end, and that one must hold to the thought that we shall soon arrive in New York. At 9 P.M. the engine stopped for the second time, seemingly plagued by other troubles which had to be determined. This time it was the helm steam works that was out of order, and it was about two hours before this was back in working condition. I was not able to understand how this ship could hold out during that terrible storm, because it seemed that everything was rusted, cracked, or full of holes. Much of the gear was half fastened down, half loose, so that there were clattering and rattling wherever one touched anything. Many among us join with me in wishing that we would soon come ashore.

## February 24

— To pass the time we were given a good task, though it was not pleasant. We were ordered (each and every one, without respect to person) to clean the ship and bail out all the trash and filth that the reindeer have left. This work was in no way ours to do, but rather that of the ship's crew. Perhaps there was some bribery between the captain and Dr. Jackson for a few dollars. I would not say that such a negotiation was certain, but I could easily believe it.

## February 25

— We had fog all day. Two west-bound steamships passed us, and after a moment they disappeared in the fog. This was the only noteworthy occurrence of the day.

## February 26

— The ship's doctor discovered illness aboard among some of the Lapp children, though of no serious nature. Everyone had to show the doctor that they had been vaccinated, for which he gave each one a certificate of vaccination. Kjellmann gave the order that everyone should pack his belongings together and be ready for landing in two days. There was activity and joy among us, over the fact that we should soon leave the bad conditions of the *Manitoban*. We all knew that things could certainly be no worse for us, wherever we went from here; so there was no more eating or drinking. Our only thought was to get off the ship. At 8 P.M. we passed the first lightship, which lies 100 miles from shore.

## February 27

— Sunday. Everyone was up early, for this is the day when we should first see land. The weather was clear and the sun shone warmly, so all climbed to the main deck in order to observe our first sighting of land on the horizon. All are gazing forward, as is usual when a ship comes in from the sea and everyone wants to see land ahead. At 10 in the morning, land was sighted to starboard, appearing as a fine strip with small elevations, showing through the fog. We could see many ships maneuvering along the coast, and we were certain that we should soon change our position. In the afternoon we passed a suburb, a town that lies outside of New York, and there we picked up a pilot. It was past dark and late in the evening before we reached the harbor. When our ship glided into the fjord, it was interesting to see the myriad of lights all along the coast; it seemed like a sea of fire, illuminating the entire fjord. At 8 P.M. we anchored in the harbor just off the fortress and we lay there all night, until morning, after the customs inspectors had been on board. As soon as the ship had dropped anchor, a number of small boats came from both sides of the harbor and surrounded our big ship on all sides. When we asked what was going on, we learned that they were sent out from large newspapers, to get stories for their morning editions. Some wished to come aboard our ship, but the captain sent them away. All of us went to rest then, in order to be up early the next morning and to see the great city of New York.

## February 28

— When morning came, I went to the main deck, looked about, and saw that we lay where the ship had anchored the past evening. We were out in the harbor with high battlements on both sides,

and some with countless cannon ports pointed in all directions. There were many large buildings ashore with smokestacks and chimneys, which I assumed to be factories. Between buildings I noted how locomotives flew in all directions. In the harbor there was even more activity. I thought that everything in the harbor would melt into each other: ships went in and out, in all directions; large and small paddle-wheeled ships whistled and tooted so loudly that it was impossible to hear what anyone said. I could not see a single rowboat, for even the smallest boat went by steam and flew forth and back in the crowded harbor along with several hundred others.

Aboard the *Manitoban*, in the harbor of New York, 28 February 1898, members of the expedition wait to go ashore.

—*Photograph published in* Frank Leslie's Illustrated Newspaper, *17 March 1898.*

After the customs inspectors had been on board, anchor was weighed and our ship proceeded toward a wharf for unloading. Along with us into the harbor went the steamship *Thingvalla*—the ship that rammed and sank the *Geiser* on the Newfoundland Banks. I immediately recalled the sad song that has been sung everywhere in Norway about that collision, in which many people were killed who were on their way from America back to their native land, which they never saw again.

The entire Lapp-Norwegian hodgepodge stood on deck and tried as best we could to open our eyes wide enough to take in all the sights. Everything was completely new to us. As the ship glided along, we stared from one side to the other. Kjellmann also came on deck and called our attention to one thing—a large human figure standing on a skerry or stone foundation, directly ahead of our ship at a distance of about one mile, and he explained to us what it was and what it signified. The figure was large, we could all see that; but I did not learn its specific size. Kjellmann said that if a grown man went up and stood on the teeth of this figure, he could not stretch his fingertips up to reach the upper jaw. France had given this to North America's United States, and it was called "The Statue of Liberty." As soon as anyone had passed this statue, to come inside the United States, he was in liberty (in freedom). When Kjellmann finished his explanation of the statue, he turned to the other side and showed us the tallest building in New York, which was 22 stories high. We passed quite near that building and were able to examine it quite closely. Farther away we saw a large bridge suspended in the air over a sound. A railway went over it, as well as a road and a foot bridge. It is presumably one of the world's largest bridges. A few minutes later the ship docked at a landing wharf, where the unloading of the reindeer began immediately. There was no shortage of spectators, as several hundred people had gathered to see the animals and the Lapps even before the ship reached the pier. Our job was to get the reindeer out, and it did not take long at all before we had the entire herd ashore. We were given our own quarters in the railway cars, where we were to prepare our own meals. Wilhelm Basi, Jeremias Abrahamsen and I became the cooks for the first car, and Hjalmar Hansen, Lauritz Stefansen and Johan Losvar were the cooks for the second car. We had no time to cook or to eat, however, because there were so many people gathered about in the afternoon. We had to observe and study all of these people. Many who spoke Norwegian came and talked with us, but there were also many English-speaking people who came and talked to us. We could not understand what they said or what they asked us. We could only shake our heads each time they asked us anything, although we could tell that their questions were always about the reindeer. Many were so ignorant that they asked how many of the animals had calved on the Atlantic Ocean, although those who had read anything about the expedition must have known that we had only bull reindeer on the ship. When we became more organized and all of the animals had been brought ashore, the first thing that the Lapps had to have was a bit of whiskey (American hard liquor), after their long journey across the sea.

## March 1

— I thought it was lively yesterday, but on this day it was ten times more so, because thousands of people flocked in to look at our company—something they had never before seen. I counted no less than twenty constables who went around the place where we were in order to keep the throngs of people at a distance. Nor could anyone walk about town in Lapp costume, for that would arouse attention and people would cluster about the individual and perhaps trample him. It was curious to observe how people here became upset when they, English-speaking, came to talk with us and we could not understand them, although we realized that they were asking various things about the trip and about Norway and Lapland. Each time they spoke, they most often received a shake of the head and had to go on to the next man, where they received the same sign. The only ones among us who understand English are Thoralf Kjeldsberg from Kaafjord and Jafet Lindeberg from Kvenangen.

The reindeer were divided into two groups, as were ourselves, and each group was assigned its own train. The first train departed from New York at 6:30 P.M. When the train began to move, people large and small alike began on both sides to shout a sort of "Hurrah!" For the first time on a train, we had a lot of difficulty, both in looking out at the people and the city's appearance, and in observing what would happen when the train began to put on speed. I thought at first that our train would tip over. We did not get to see as much of the city as we wished, since the railway station where we boarded was at the edge of the city. After only a few moments our train was headed towards the West Coast at breakneck speed, while the next train was to depart half an hour later.

## March 2

— The first thing I did in the morning when I awoke was to get dressed as quickly as possible and run to the window in order to see how the countryside looked. Everything seemed to me to be exceedingly beautiful, especially the landscape with plains, small hills, forest, and without snow. At times, I did not know when we were in cities and when we were not, since it was built up and inhabited everywhere as far as the eye could see. I therefore assumed that it must be one continuous city. During the day our train stopped in a large attractive city called Pittsburgh, where a multitude of people stood by. I felt certain that there were several thousand who had thronged together at the station to see the Reindeer Expedition, which a telegram from New York had told about. There was no lack of audacity, either; for if they could not come into the cars to see us, they crept up to the windows in order to peek through the glass into our quarters. Some who could not get close beckoned to us, indicating that we should come down from the cars so that they could see and talk to us. The women in particular seemed persistent; and if one of us sprang down from the cars with a Lapp cap on his head, a number of women came to seize upon him. Not only that, but they gave their cards to the boys, saying that they should write to them, married or unmarried. Some wanted to have locks of hair from several of the Lapps, and others wanted to have a bit of reindeer moss as a curiosity. Not only in Pittsburgh were people flocked together to see us, but after

Men from Finnmark disembarking the reindeer from the *Manitoban* at Jersey City, 28 February 1898.

*—Photograph published in* Frank Leslie's Illustrated Newspaper, *17 March 1898.*

we left New York they were at all places where the train stopped, whether it was day or night. We who were the cooks on the train had little enthusiasm to keep at food preparation, since we, too, wanted to see everything noteworthy, and there was something to see every moment.

Nor was there any lack of amusement each time the train stopped. One could find out how many minutes the train would be in station, and thereupon jump out into the multitude of people. Then would be a lot of activity, for all wanted to see the foreign Laplander, whose skin was just as white as their own, although it seemed that they had thought that the Lapps' skin was as dark as that of the Indians. They were especially astonished at the Lapps' clothing, which was so incomprehensibly strange to them. I had a pair of Lapp moccasins on my feet and a fine Lapp cap on my head; a number of women wanted to see my cap so I took it off my head and extended it to them. I had to laugh to see how they handled my cap, for I could never have handled a small bird, or an egg, so carefully as they handled my Lapp cap. Then they saw the moccasins, which I had on my feet, and they were so careful about touching them that I could not feel a thing. They shook their heads and spoke to one another in English and seemed to be quite enthralled.

After we had gone to bed, we were disturbed by an uproar outside each time the train stopped at a station; and I observed how one person would help another climb up to a window in order to peek into the cars, perhaps to see something strange; but we were all in bed asleep, and besides that, it was so dark that they could not see anything remarkable at all.

## March 3

— At 5:30 in the morning we came to Chicago, where the train stopped for three hours. As we arrived so early this morning, not a single person was at the station; we did not get to look around in the great metropolis since the train stopped at a less distinguished place, on the edge of the city. We were not able to leave the cars because time was too short for us to walk around the town; we might be left behind. At exactly 8:30 the train departed, and not long afterwards we stopped in a lovely town called Milwaukee, perhaps the most beautiful of all the towns we have passed through, in my opinion. A long line of people stood along the railroad track in order to see the reindeer and the Lapps, as the train stopped for a few minutes. I was told that people had been standing there all morning without having taken time to eat breakfast, so that they would not be crowded away from their places when the train came. A lot of children came running up, some large and some so small as to barely walk. They stood there and froze until they were blue in the faces. But when the train began to move, the children shouted "Hurrah" at the tops of their lungs while they stood there staring after us; possibly they would wait for the next train, which was only a half-hour behind us.

## March 4

— The day began as the others, by my looking in all directions. Each time we passed a station, people were gathered together out of curiosity. I get the impression that Alaska must be frequently talked about. Children shouted "Klondike" time after time; and many people who spoke Norwegian told us that we were going to Alaska and could find gold and get rich; so there were many who envied us this journey. In the afternoon we approached the great prairies; their expanses were so wide that we had the horizon to all sides. Here and there a number of houses stood clustered together, but we could travel several hundred miles without seeing other than uninhabited plains and forest of some sort.

## March 5

— At dinner time we passed an Indian village, which was new and interesting for many of us. While the train was stopped there for a moment, we had occasion to observe these sun-burned people. Nor were they slow to run to the train cars to look at us, since they had been told that a number of Indians had come on the train from Scandinavia, from the frigid climes, and that they were of the same tribe. So while the train was stopped, it was a matter of staring at each other, eye to eye, considering each other's appearance, and color in particular; and clothing, which was quite different. Most of the Indians had only a blanket or sheet of red material, and this was thrown over the shoulders like a full-length gown covering the whole body, and there were some who had no other clothing on at all. Their faces were dark; a number were copper-colored and many I presumed had painted themselves, so there were all sorts of facial colors to be seen.

On these great plains the trains always travel at a greater speed than through cities and towns, since seldom or never are people on the tracks. We were told that our train traveled at seventy miles an hour.

## March 6

— Sunday. Previously we have seen great plains; today we get to see mountains, hills and forests. The trees were thick and several hundred feet in height. They were so close together that one could hardly see more than a distance of twenty feet through the forest. Most all day we traveled thus, until we came out of the mountains in the afternoon. In the mountains we passed through many tunnels of all sizes, where it was dark as the grave for several minutes, although the train went at full speed.

## March 7

— Today we encountered a high mountain that the train had to climb, and when it reached the top it had to go down on the other side, equally steep. There one could see other trains ahead of us, how they tottered on the mountain side in descending. Ascent and descent took several hours. Here we saw our first snow since we left Norway (except for ice on a number of rivers and frost on the ground in Wisconsin and North Dakota). Then abruptly we came out of the mountains, and just as our train flew down a valley where the rails lay well anchored to the ground because of the river that flowed right beside us, one of the cars jumped from the rails. The whole train received such a jolt that we were nearly injured; but luckily the car jumped to the left, against the hillside. Had it jumped to the right, the entire train would certainly have rolled down into the river and crushed everything, both living and dead, and buried us there. They had the cars back on the tracks in an hour's time and proceeded onward, although the engineer was injured when the car jumped, and his face was bloody. The reason for the car's jumping the tracks was that a stone had rolled down the hillside and stopped right in front of the wheels, causing the car to lurch to one side.

Thus we traveled across the rest of the country and down towards the coast, through farmers' properties where the meadows stood green and near to bloom. Late in the evening our train stopped on a long bridge about seven miles from the city of Seattle, which lies on the Pacific Coast and is one of the largest shipping points for Alaska, on the coast. We lay there all night on the bridge, and not until morning were we allowed to come into town and change our mode of living. Then we were almost exhausted by the continual riding on the train. Although we did not know what to expect, many hoped that it would become better and more

MEN EMPLOYED BY THE GOVERNMENT IN CONNECTION WITH THE LAPLAND REINDEER EXPEDITION.
Photograph by La Roche. (See pages 39, 105, and 106.)

Personnel of the Reindeer Expedition, in Seattle, 1898. Tentative identifications of four: Top row, sixth from left, Jafet Lindeberg; next to him to the right, William Kjellmann. Top row, second from far right, Carl Johan (in military jacket). Second row from top, third from left, Wilhelm Basi.

*Photograph by La Roche, published in Sheldon Jackson's report, 1898b.*

comfortable the farther we traveled, as one soon becomes weary and desires a change. Train No. 2 caught up with the first train in the evening and spent the night on the same bridge.

## March 8

— The trains took us about four miles closer to the city, to the edge of a town which was called Fremont, where there was a large area known as Woodland Park. All the reindeer were let loose outside of the park gates, and we get to stay in the park building, which stands in the middle of the meadow there, with tall, lovely trees roundabout. When we arrived at the park, all of the cooks were given a vacation, and everyone had to be responsible for preparing his own food.

## March 9

— People now knew that the reindeer and the Lapps were in Woodland Park, and there was no shortage of guests. From early morning until late in the evening people came and went, from both city and countryside, to see Lapps and reindeer—such they had never before seen. Many Norwegians came who spoke with us, and among others there I met one of my neighbor lads from Norway who had been in America for twelve years. His name was Hedley Redmyer; he was hired as an interpreter for the expedition. In the evening a number of Lapps came home from town where they had gotten hold of liquor, whereby they created a racket all night long.

## March 10

— Since the Lapps had abused their liberty last night they were not allowed to go out of the park all day, to keep them from getting liquor. In the afternoon, Wilhelm Basi and I received permission from Kjellmann to take a trip into town to look around. We asked Hedley Redmyer to go along and trudged to Fremont where we took the trolley car, which carried us straight into town. The trolley tickets cost five cents per man, and for that we got to travel about three or four miles. Arriving in town, we tried to observe all curiosities. Mostly we stopped and looked around in the enormous general stores where there were all sorts of things of all varieties for Alaskan gold miners. For sale for Alaska were articles of all models and styles, fashioned as light and easy to handle as possible. Life on the streets was generally bustling, and all of the main streets were full of people. I did not know where we were going; I thought that the city was large, with a population estimated at about 100,000. When we had had our fill of walking—and we were tired, too—we then went up into one of the trolleys. I did not know where it would take us, but we followed our companion Hedley Redmyer. We paid another five cents and after we had ridden for a little while I knew where we were. We quickly got off at Fremont and came up to the park where we were staying in only a few minutes' walk.

## March 11

— The Lapps received permission to travel to town. Many people were looking around in the park, including many Norwegians, Swedes and Danes who could speak with us; there were otherwise people from every nation. In the evening, when the Lapps came home from town drunk and wild, we got Samuel Balto from Karasjok to dress up in Norwegian clothes to pose as a policeman. He put on his head a visored cap with a large, silver medallion on it. When Balto came forth dressed in this manner, the Lapps believed (as we had told them) that it was an American policeman who had been sent up to the park in order to keep the peace. Then, those who were drunk became quite peaceful.

## March 12

— It was quite peaceful in the park, in comparison to the previous days. Only a few strangers walked about and spent time inside the park gates, where the reindeer went during the daytime.

## March 13

— Sunday. Today was a visitors' day, and the whole park was full of people who had come from town to look about. Among other things, there were a number of fine ladies, who we thought were all too bold. One Norwegian lad who was being harrassed took the young lady in question and kissed her while the entire crowd stood and watched; everyone laughed at this, but the girl laughed as well, as if nothing had happened. The next day we could read in the newspapers how the ladies of Seattle had amused themselves up in the park on Sunday with the Lapp boys.

## March 14

— We received orders to be ready by tomorrow to put all of the animals into the railroad cars that would take us to Seattle, to the shipping-out area for Alaska.

## March 15

— Early in the morning we began to move the reindeer down to the cars, the job lasting more than half the day. A little Lapp boy who had been ill during the entire journey died during the morning. The Lapps sang over the body, as was their custom. Since we all had to leave shortly, the body was turned over to a funeral parlor. We then said farewell to the park and journeyed to Seattle, where we were shown aboard a large three-masted barque that lay by the wharf. All of the reindeer were put aboard; and we got to eat our lunch ashore while the ship lay there.

## March 16

— After we came to Seattle, we no longer have Kjellmann as our leader as before. The leadership has been turned over to a group of officers who knew neither up nor down, and everything stood still under their command; nor could they get the reindeer aboard the barque, but they had to get Kjellmann to lend them a hand, as usual. Everything went well when Kjellmann was at the helm.

## March 17

— During the night, while we were sleeping, our ship left Seattle, towed by a tugboat. In the morning we stopped at a little town called Port Townsend, where all Lapps who had families were put ashore. We later learned that they were to stay there temporarily in order to take steamships to northern Alaska when navigation channels were open, while we were to go to Dyea and walk from there with the reindeer across the mountains to Circle City. When all the women and children had been set ashore, the boat took us in tow again and set sail northward, through beautiful, narrow channels. At dinnertime we had the first chance to see how the cooks appeared and how the food tasted. To our discomfort, a Negro and a Japanese stood by the stove and stirred two large kettles of stew, potatoes, and some sort of canned meat, all cooked together. We felt that we were once again on the Atlantic, and therewith we were upset, although we were not quite so badly off as we had been, for we experienced no storm nor heavy seas, as we sailed in coastal waters through narrow channels—a desolate but beautiful landscape, forested everywhere. We had ample time to

The steam tug *Sea Lion*. *Photograph courtesy of the University of Washington Libraries, Special Collections, Negative 8417.*

observe our surroundings in the warm sunshine, going at a leisurely pace, as our ship made only about five miles an hour and sometimes less, when the current shifted against us.

## March 18

— Most everything has been quite good. But there has been constant quarreling with the two cooks, as they have been apathetic in their handling of the food—which we have constantly noticed. We had to call Kjellmann back down to the cabin to attempt perhaps to straighten things out. It went to such lengths that the cooks said that they could not manage more and that they wanted to have one or two men to help them, in order to carry out their work properly and cleanly. They would have had ten men, had they been available. But they required only two, which work fell to Wilhelm Basi and Jeremias Abrahamsen. But this did not seem to change things noticeably.

## March 19

— We got a head wind. The tugboat could not manage to go against it with the great barque and had to run before the wind down to a bay in order to wait out the storm. Along the shore there was nothing but trees that had been twisted up by the sea, and otherwise not a living thing. We lay there in the bay until the next morning while the storm blew out in the channel.

## March 20

— Sunday. The main thing that was different as we came farther north was the greater number of mountains. Although some were rather high, the forest lay like a sheet all the way over their tops. It seemed to become a bit cooler, too, the closer we come to Alaska.

## March 21

— We have traveled in coastal waters the entire time since we left Seattle. Today for the first time we crossed a bay thirty miles wide. When we had sailed over it, we came into a strait where the passageways were so narrow and cramped that one could easily have leaped to shore from the ship. After dinner, we went by Charlotte Harbor, and a very lovely little town on a narrow channel. We stood then gazing about and walking about the deck for the rest of the day.

## March 22

— Mild weather with rain, so there was little difference in weather between Seattle and the southern coast of Alaska.

## March 23

— We lay still the whole day, due to snowstorm and fog. We have not experienced a snowstorm from the time we left Norway until now, on our journey to Dyea, Alaska.

## March 24

— In the morning the weather lifted, so we had another change by traveling farther. We crossed the border from British Columbia into Alaska Territory, where at the customs station all ships must put in, both north- and southbound. In the evening a few of the fellows cut loose in the galley in the darkness, just for fun, and stole a whole tub of butter. There was never enough butter at the table, and now and then, none at all would appear, as the fat, corpulent supply master and the two cooks were too stingy with it.

## March 25

— Now we had adequate butter; even the Lapps' dogs got some. Neither the cooks nor the supply master had any knowledge of this. At dinnertime, a ship passed us northbound, headed for Dyea with passengers. Since that vessel had to go nearly alongside our ship because the sound was so narrow and shallow on both sides, we could recognize Dr. Sheldon Jackson, who stood on the bridge looking over at us and waving with his handkerchief, whereupon we immediately responded by shouting "Hurrah," led by Kjellmann, who was with us. We were answered with a doubly loud "Hurrah" from the steamship, which glided swiftly by us and was out of sight in a short time.

## March 26

— A cold northwesterly wind blew with driving snow; the day was not so pleasant, since the ship had to lie stationary for a long time due to the storm.

## March 27

— Sunday. We left early in the morning. The air was still and clear, with sunshine, as we glided into the fjord toward Dyea, and at 4 o'clock in the afternoon we anchored at Haines Mission, just a few miles outside of the town of Dyea, where we were to go ashore with all of the reindeer. From there we are to begin on the journey over the mountains to inner Alaska. The terrain at Haines Mission looked beautiful, with high, forested hills. Down by the shore, it was level, but the snow lay everywhere on the ground, to a depth of several feet. A large number of white tents of all sorts stood on the sandy beach. Some were round, others semicircular, and others were rectangular like a house. It was said that this was a camp of the soldiers who had recently arrived and who were going with us up to Dawson City in inner Alaska. Farther away stood a cluster of houses—the winter huts of Indians. Just beside them on the hill was a lovely large, red-painted building, which was the mission house or school. Close by were the parsonage and a small hotel for travelers to stop over. This was all I could see at Haines Mission, where we are to be set ashore in the morning.

## March 28

— The day began with sunshine and calm. Aboard our ship it came to be anything but calm, because an investigation by Kjellmann found that there was very little reindeer moss left. This

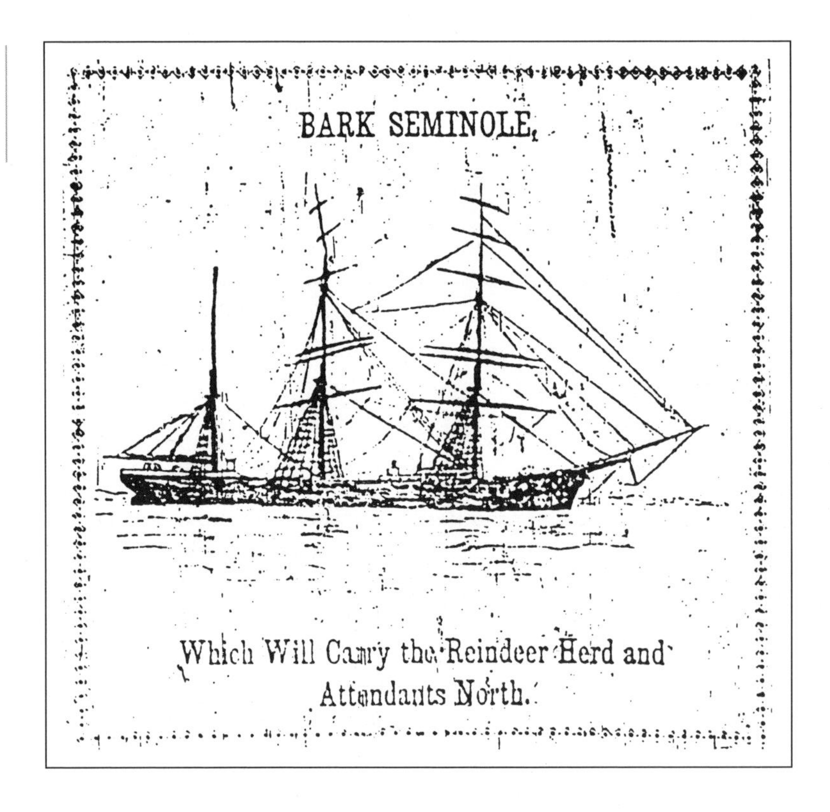


The *Seminole*, as illustrated (artist's sketch) in the *Seattle Post-Intelligencer,* 15 March 1898.

came as such a surprise to Kjellmann that he could only remain silent, for long periods of time; [we thought that] the shortage of reindeer food would soon be felt quite desperately. The reindeer were then brought ashore, while Kjellmann arranged for four from our group and one soldier to go into the mountains in order to find moss. They left immediately in the morning while we others were getting the reindeer ashore, which operation we finished in the evening. Two men were put on night watch with the herd while all of the others went back on board for the night. Seventeen reindeer had died out of the original herd of 538, most of those in Woodland Park; we landed at Haines with 521 animals.

## March 29

— We moved ashore in the morning after breakfast, and Kjellmann went ahead in order to show us into a house where we should live while we resided in the flatland by the sea. When Kjellmann headed towards the parsonage, we thought that we were to have one of the rooms in the parsonage or otherwise in a division of the mission house; but we received neither of these and had to be content with the pastor's barn, amongst the cattle and horses. For meals, we had to walk half a mile to a little wooden hut where the Negro boy prepared food for us, cooking the food outside as there was no room for that in the building. It rained quite hard in the afternoon, and the unfortunate reindeer stood tied inside a fence right by the mission house without food, because there was no more reindeer moss. Everything had been used on the journey on the ships. All that the animals got was a sort of coarse straw-hay, which the cattle could not eat, let alone the reindeer, who are not used to eating much anything but moss. No other food was given to them, and they could not eat the hay that was brought in to them; so it was a matter of either life or death. At night two men were put on watch with loaded guns lest dogs should approach the herd, in which case the watch had permission to shoot them.

## March 30

— It rained all night and seemed to keep the daylight away, and in fact, it did. We had a chance to look at the Indians' fishing gear, which was most remarkable; I cannot comprehend just how they could catch fish with their home-made hooks, which were of both wood and metal, but mostly of wood. A hook resembled an ordinary compass with a number of fish figures carved on it. Then a nail about three inches long was driven through one part of the compass at an angle, forming a half-inch opening to the other side

Carl Johan's sketch (original and enlargement) of the fishhook of the Tlingit people, at Haines Mission, 30 March 1898.

of the compass, as in the figure, with carvings on all sides. With these they fished the large halibut and came rowing ashore in their peculiar boats, which were made completely of wood in such a manner that there was neither plank nor hide on them. They were simply carved and dug out of great, thick logs, some of which were from twenty to thirty feet long. The large boats could be six or seven feet wide. In the evening, we received permission to go into the mission house to listen to their prayer meeting and singing. The pastor preached in English, and his words were translated into Indian language by a lovely Indian girl. The meeting concluded with singing and a closing prayer.

## March 31

— I saw something very unusual, compared with our custom in Norway, when the pastor himself went into the barn and tended to the animals and milked them, too. We had good occasion to observe that, as we were camped right in among the animals. In the evening, Dr. Jackson sent a message to the barn to come to the evening meeting in the mission house, and we all went. Dr. Jackson gave a talk, which was translated by the Indian girl into Indian language. Then Dr. Jackson addressed us directly, and his words were translated by Wilhelm Kjellmann. Next, we sang a song—the well-known "How blessed is the little flock," which everyone knew. Afterwards, the Indian girl and the children sang in English, which was lovely to hear, for the Indian girl had an exceptionally good singing voice. Dr. Jackson said a prayer, whereupon the meeting ended.

## April 1

— The men who went to the mountains in search of moss came back without having found such food for the reindeer; they had now been without food for several days, bound to the fence in rain and storm, as there was no other recourse. A new attempt to find moss was duly undertaken, also without success. So there stand Kjellmann and Jackson and the rest of us beside them, completely helpless, while the unfortunate animals stand and starve.

## April 2

— During speculations as to what we could do in order to save the animals, and since we had nothing else to do, Dr. Jackson sent a message up to the barn in the evening that we would meet in the mission house in the schoolroom, as he wished to deliver a sermon. We all were willing to go and we met after dinner at the school, which had been requested for our use. Dr. Jackson gave a talk especially for the crew of the expedition, with Kjellmann as interpreter. His talk began with exhortations to God that we should try to lead a God-fearing life, always wisely, and each go forth as a good example, for we might in many places come together with people who had never heard of God or of civilization; that we should also be as a light to them, which should be a reward and a joy to us; and that finally we should win the praise of people where we journeyed, and they would speak well of us; whereby at the same time, we should give him a great joy if we lived in such a manner. Dr. Jackson then promised us aid and protection in all circumstances we might come into, saying that he would always think of us as would a father; but he would now leave us and perhaps not see us again before summer or next fall. Thus was Jackson's speech to us, and then he was going to leave and travel down to the States.

## April 3

— Palm Sunday. A great number of raincoats, rubber boots and sou'westers arrived for us to use. We had gone all of a week without rain gear in the terrible rainy weather and never got our clothes dry, as there was no stove in the dark and dirty barn. It had rained incessantly, night and day since we arrived.

## April 4

— The day went as usual at a standstill. Kjellmann issued orders for tomorrow to be ready to set the reindeer free and to try to get them out onto a promontory about six or seven miles from the mission. At the promontory the snow had been melted away by rain and storm, so we would try to get the animals over there—perhaps there was something nourishing for them that they could find when they went about on their own. Something had to be done. The reindeer could not hold out much longer, as they had now withstood several days without food, only the dry, coarse straw-hay being all they received for nourishment.

## April 5

— At 4 o'clock in the morning we began to move the herd out onto the promontory. We had to get the herd over an isthmus to the other side, where there is an Indian village called Chilkot. The nearby river, which flows down into the sea, has the same name, Chilkot River. Then we drove the herd to the headland which extends far out towards the sea along the beach strand. But the animals could not manage to go farther, due to their hunger, and we had to stop and permit them to graze in a small meadow. But no fresh grass was present, since the snow had just melted away; only raw, dried grass and a little moss were on the mountain slopes where it was difficult for the animals to reach. Although we had moved only about seven or eight miles, many of the reindeer lay exhausted along the shore, and a few that could not walk any farther died there. Kjellmann then ordered some men to stay and watch over the herd, while the rest of us returned to the mission.

## April 6

— For those of us who returned to our beloved barn, Kjellmann had arranged that we no longer needed to walk around the pastor's stable area to keep an eye on it. We had our hands full of work—even more than we could manage—in pulling the reindeer sleighs and supplies over to the river, about one mile from the barn that was our headquarters. We heard today of a tragic drama from Skagway, near Dyea, where 265 men were buried by a snowslide. All were goldminers who had camped beneath a cliff, from where the avalanche loosened and covered them all. Only a few were rescued.

## April 7

— Maundy Thursday. We were divided into four groups, each with a foreman. We did not know who was going to the mountains, but we assumed that no one would avoid it. For that matter, everything would depend on how the reindeer would adjust: whether they died and were thus lost, or whether we would be able to find food for them in order that they would live and take us to the mountains. Many officers were waiting for a reindeer ride to the mountains along with us. Now we had to get supplies and sleighs and all of the necessary articles about fifty miles up the Chilkot [Chilkat] River. That was not to be an easy task, for all of this was to be done by manpower, as the snow was nearly all melted away, and the river was open. Kjellmann hired seven Indians, each with his own boat, to meet us at the river bank where we had our supplies, in order to load the boats and thereby travel up the river.

## April 8

— Good Friday. We did not get to observe Good Friday, for we were up at l o'clock in the morning, in the cold and rain, to begin the journey up the river with the goods that were to go into the mountains. When we arrived at the river, Kjellmann sent two men over to tell the Indians that they should come immediately with their boats, since there was high water early in the morning. But those gentlemen sent word back that they would not come before 12 o'clock. We had not reckoned that we should have to wait, standing out in the rain, and it was also quite cold. There was nothing for us to do but to pull a large boat in amongst the trees to keep ourselves warm. At the same place we had our dinner, which was of a sort of canned meat. We opened one can. About four pounds of meat are in a can, and we divided this among twelve men. In order to have coffee, we had to cook it in the same meat can (we had no coffeepot). With the coffee, we had some crackers—a sort of hard cracker with neither taste nor smell—and this was our dinner. Nor was any more than that to be had, since a ration had been measured out for each month. The ration consisted mainly of flour and pork. Thus, we rested half of Good

Friday, but after 12 o'clock the Indians came with their canoes, and these boats were loaded up within a few minutes. The largest canoe held about 700 pounds, and four men to head up the river. At times we sailed, and at times we dragged the canoes against the current. Then, some of the men (all who were to go overland) would shoulder their loads, with long and heavy rubber boots on our feet, a long oilskin coat, and a sou'wester on the head. Thus we marched along the river. Kjellmann was with us, also on foot. We walked and climbed with all our strength. At times we went over hills and mountains and through the confoundedly dense forest, where we stumbled both forwards and backwards. If one was delayed in the least, one stood a chance of losing contact with comrades who were hidden in the bushes ahead. Constantly we came upon great marshes, and had to make long detours in order to get around them. We walked all the rest of the day until at last in the evening we stopped to camp. We could go no farther because it was growing dark, and all of us were tired and hungry. Some had their clothes torn and their hands swollen from thorns. Many places where we had to go were mountain slopes, and the thorny growths stood so thick that we did not know how to protect ourselves. To fall was dangerous for hands and faces equally, and the thorns were so sharp that they cut even our feet. The people in the boats did not have it much better, however. In the first place, the river runs in countless curves, which make the ascent very difficult. In addition, many falls and treacherous places where loaded boats cannot pass are in the stream; rather, one must get out of the boat and drag it. Finally, there are some places where they had to unload the boats, cross over the sand banks with half of the load, and later go back down after the rest of it. We traveled in these ways about ten miles up the river, where we cooked up a little grain soup with a piece of meat in it which was supposed to fortify it, so that it was not simply flour and water. Right after we ate the soup there was coffee, and then we went to sleep. Kjellmann was with us under the same conditions and shared in both good and bad, although he certainly did not have to do so; but he wanted to be with us. No one was to do anything for him, and he was trying to do everything possible to get the reindeer over the mountains alive.

## April 9

— Early in morning, we continued onward as in the day before, but the route was ten times more difficult. The valley became narrower, as mountain and rock sloped right down to the river's edge in many places. We who were on foot then had to wade across the river. In order not to get their stockings and pants wet, some would take them off beforehand, and when they had reached the other side, they would put on dry stockings. Jeremias Abrahamsen had to change the straw in his reindeer-hide boots three times, after having to cross portions of the river (three times) in the cold spring water. None of us had dry feet, however; for if we did not have to cross the river, then we had the endless forest, and the swamps to go around. We did not eat all this day; the supplies were in the canoes and it was impossible to know where the canoes were, as the river bent into curves and divided itself into branches everywhere in the immense forest. We could only march along, and we knew that we should arrive at an Indian village, called Klukwan, by evening before dark. We came there at 4 o'clock in the afternoon, having walked fifteen miles. The canoes arrived shortly afterwards. Kjellmann rented two large Indian houses for the night, where we were to sleep. The houses were built of a type of boards or thick slabs of wood, bound together with some sort of root instead of nails. The roof was built on posts which rested on the rafters. In the middle of the roof was a large hole where the smoke went, from the hearth, which was placed at the middle of the floor. By the walls in each corner stood large, carved, wooden monkey-like figures, painted in several colors. Behind the figures were sleeping places on platforms. Chests of all sizes and colors, old and new, were piled up around the walls. When we had the opportunity to peek, out of curiosity, into one of the chests, we saw to our horror a human skeleton within it. We put the latch back on and let it stand in the same place. Later we learned that there were only coffins in the houses; there was a goodly stench as well. The Indians came around where we were in order to see especially the Lapps, who aroused their attention. They looked at the Lapps with admiration, and at the Lapp knives as well; all of us walked about with large Lapp knives in our belts. I was assigned to cook for a group of the Lapps. After evening drew on and we had gone inside, I made some pancakes and prepared soup and coffee. Jeremias Abrahamsen, who was cook for the Norwegian division (with twelve men in each group), did the same. The third cook, Johan Losvar, was with the herd and cooked for the herdsmen.

## April 10

— Easter Sunday. Inasmuch as we had thus abused Good Friday, Kjellmann gave us permission to rest on Easter Sunday. So we spent Easter in the Indian village. The only entertainment we had came when we asked the Indians to show us their dancing—and it was interesting but unpleasant at the same time. They made all sorts of twistings and contortions with body and eyes, and their

Opposite: Southeastern Alaska. The route of the *Seminole* via Lynn Canal to Haines Mission. The members of the expedition who were ordered to return to Haines Mission departed from their friends near the headwaters of the Klehini River—a locality indicated by the black bar and arrow. The surviving deer were led during the following weeks through Chilkat Pass and towards the Yukon River on a course as shown (approximately) by the broken line to the north of the pass.

singing during the dance was terrible to listen to. Jafet Lindeberg, who had asked them to dance, then in return had to show them how white people dance, and Ole Rapp and he danced a Reinlender for the Indians. There in the simple, sad Indian village, we held our Easter, inside a filled-up burial house, and the day finally seemed as long as ten; but we were obliged to be there for only one night, and we tried to make the time pass as easily as possible.

## April 11

— In the morning, we threw the gear into the canoes and traveled farther up the river, in total about two miles, to the stopping place where we were to unload our goods. We were there in a few hours, and there fifteen men were left in order to haul the supplies on sleds farther towards the mountains, while all of us others returned to fetch the reindeer and the reindeer sleighs. When we went down the river with the stream, it took only a few hours before we were again by the sea.

Along the shore there were great piles of dead salmon that had become frozen in the ice during the winter. For a long distance, the smell was like a whale factory. We arrived at the mission after dinner, whereupon we ran, as the fox springs into its hole, into the black barn. The barn was indeed a welcome sight, for we had been out for several days without much protection from the rain, which had been falling night and day.

## April 12

— Kjellmann sent two men to the herdsmen on the promontory with word to hold the herd together, in order to attempt to bring them up to the river the following day, where we waited with the boats, reindeer sleighs, and supplies that we had brought over from the mission. Instead of going home (to our barn), since it was late in the evening and we were tired, we went into an Indian hut. Kjellmann was also with us; since the door was closed, he opened a window, and through it we all crept in; there we prepared our usual evening meal of coffee and dried crackers. Thereupon we went to bed, wet and somewhat miserable. Each man's bedding was a wet oilskin coat and a wet sweater over the head. The beds consisted of some pieces of wood, which the Indians had used as a floor. There was storm and rain and the hut leaked everywhere, so we lay in a clump right in the middle of the hut, but not sleeping. First of all, it was cold; then there was the rain. When the storm was at its worst, it seemed about to overturn the entire hut as we lay inside.

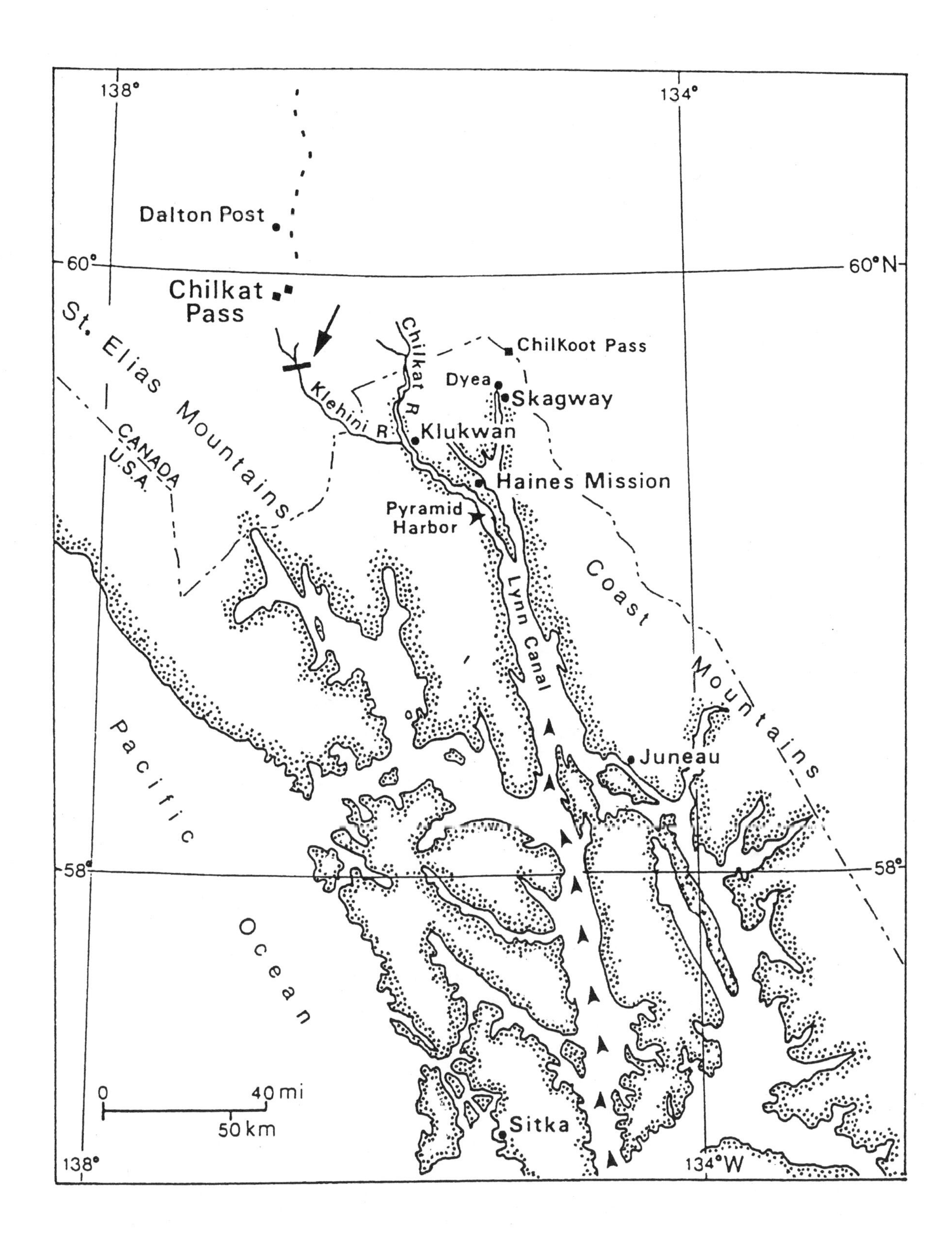
138°
134°
Dalton Post
60°
60°N
Chilkat Pass
St. Elias Mountains
Chilkat R
Chilkoot Pass
Dyea
Skagway
Klehini R
Klukwan
Haines Mission
CANADA
U.S.A.
Pyramid Harbor
Lynn Canal
Coast Mountains
Juneau
Pacific
58°
58°
Ocean
0
40 mi
50 km
Sitka
138°
134°W

## April 13

— We did not have much work, as we waited for the herdsmen to arrive with the reindeer. All day we waited, but neither herdsmen nor herd came. At this point, we loaded our little lifeboat with some "reindeer supplies" and some reindeer sleighs and took it a distance up the river, where we had spent the night before. Just as we arrived at the hut, a message came from the herdsmen that the entire herd was dispersed. It was spread out over the length of the great promontory, which was several miles wide and perhaps twenty miles long. The forest was so thick that it was an impossibility to get through it, especially now that the snow was so soft that one fell in deeply. Now once again, Kjellmann pondered. From where was all of this trouble coming? We were visited by one setback after another, which was leaving us tired and dejected. All of the toil and work would be in vain, although we worked with perseverance and did not hold back in the least to save the reindeer. And already two men were in the hospital, seriously ill. We lay down early in order to be up early next morning. This night began and ended the same as the previous.

## April 14

— Kjellmann sent fifteen men down to the promontory with supplies of only flour and meat. The men there had nothing to eat, as all of their supplies had been used up. I also was sent down, assigned, with Johan Losvar, to cook for the group. We did not arrive at the herdsmen's camp until the afternoon, and some of them were so hungry that they were ready to eat the cooks; so Johan Losvar and I began immediately to prepare pancakes and grain soup and coffee. Everyone received these dishes, but not as much as he wanted. When they had had some sustenance, they were to return to the forest to search for the reindeer. Some of the boys received two pancakes and a small piece of raw meat each, and having eaten that, set out without the least bit of food to have later on in the forest. They were so persistent that they kept traveling farther when evening came, rather than coming back to the camp, in order to search earnestly for the deer. They had to spend the night in the dark forest in the snow and no food at all for next morning. All night it rained, as if coming from an endless source in the sky.

## April 15

— In the morning, everyone went into the forest; voices could be heard from all directions. Some returned with two or three reindeer, but the average was ten or twelve—a number that one could manage to drive through the trees. But some fine fellows came with a whole herd, and the boys who had been gone all night returned with a fine herd. Among those were Otto Greiner, and Peder Johannesen Lakselv, and Samuel Josefsen Lakselv—but I do not know if more than those three were out over night. Enough were found that made up a herd of three hundred reindeer, and these were gathered in a little meadow by the seaside. In morning, we would take them to the river, about seven or eight miles, according to orders that Kjellmann had given (to go by the river). Late in the evening, Kjellmann came with some men from the mission who were to help when we passed by the Indian village of Chilkot, since there were so many Indian dogs and we had to go right by the Indians' homes with the herd.

## April 16

— Most of our group went towards the river, while ten men remained out on the promontory in order to try to find yet more deer. Lindeberg remained there with his group; but within one day they were to go to the river in order to come with us to the mountains. The whole day was taken up before we arrived at the place where we were to stay over until the following day, when we would try to get the reindeer across the river. The entire way along the shore from the promontory was strewn with exhausted deer. Many walked until they fell down and died on the spot. In the evening, Lindeberg came with fifty reindeer that they had found in the forest out on the promontory; but the herd did not get any larger for that, for there were nearly as many dead and ill as those he brought in. I shall not estimate the number of living until we have been able to get them into the mountains, to see then how many are still alive. Now we could also observe how many are starving and tortured; for they are so hungry that they ate the bark of spruce trees and licked each other a little, and even bit into our clothing, which was pitiful to see. Everything was done to save them, and we shall later know the results of all of our difficulty and toil.

## April 17

— Sunday. We drove the reindeer out into the river, whereupon they swam over to the other side; but although the river was no broader there than that one could easily throw a stone across it, there were many that drowned. They could not manage to swim across. The people were transported over with a small boat we had with us. We moved the rest of the deer about seven miles after we had got them across, and then it was not possible to move them farther. All of them lay down, and neither dogs nor people could get them to stir. Down along the river shore, as far as the eye could see, stood and lay reindeer that could not manage to follow us. A number toiled their way to the camping place in the evening; most of them never made it. It was not only the deer that should suffer and have an unpleasant time, for we had a sampling of the same, with so little food that no one could taste it when we looked at it. We had a drop of grain soup for dinner, after which we had to meet the long, cold night without complaining or grumbling. The reindeer now were relatively well off, as they had a sort of swamp grass to eat. Each of us with his own sack in hand climbed up into the rocks on the other side of the river and gathered what little moss that we could find, and carried it to the deer.

## April 18

— All day we worked with the same task we were assigned in the morning: We ran from one cliff to another and put everything that even resembled reindeer moss into our sacks, while Kjellmann went down to the mission with three men to get food for us, since we were completely out of supplies, too. Kjellmann left orders that we were to try to move the reindeer herd about fifteen miles farther up the river the following morning, and then we were to stop by the river bank so that they could see us when they came hiking back up the river.

## April 19

— According to Kjellmann's orders, we moved the herd, and stopped by a grassy islet near the river bank, where we bound all of the reindeer fast to the bushes, so that they could eat the grass that was beginning to turn green. That done, we all traveled by boat to the other side of the river again and went to the mountains, each with his own sack, in order to find moss. Fortunately, we were able to gather several sacks full, and a mouthful was given to each one of the animals when we returned to them. As the evening wore on, Kjellmann still had not returned. We cooked the last of the grain soup and began to watch down the river to see if we could not soon find Kjellmann. We gazed into the twilight until we finally perceived many Kjellmanns coming, but none of them ever got closer. But after dark, Kjellmann came, and with supplies, and we cooks had to make soup and prepare cakes and coffee for supper. Afterwards everything seemed well and satisfactory.

## April 20

— We moved a good distance up the river and stopped by a swampy meadow nearby where was grass that the reindeer could eat. They were not able to get anything else, since no moss was there. Then we prepared our beds of spruce and pine boughs and slept peacefully through the night under clear skies.

## April 21

— We traveled farther, each with his own sledge. We walked until we reached the Indian village of Klukwan. A large number of reindeer had fallen behind the main herd of this day, and those who came through were no more than barely alive. All they got to eat was old hay.

## April 22

— All day we lay quietly, not just for the sake of the animals that needed a rest, but because the people who had worked and labored relentlessly for several days also needed a bit of rest. Kjellmann sent two men onward in order to see how far the fifteen men had come, with the supplies that had been left above the Indian village from the previous trip. If they could find them, they were to come back as soon as possible to the place where we are.

## April 23

— In the morning the men who had been sought came to join us. They had their camp about twelve miles farther upstream. Kjellmann arranged that we should go down again to the mission in order to fetch the reindeer sleighs, while the fifteen men should take the herd farther up river. On the way down, we met Lindeberg and his men coming with the rest of the reindeer they had found. When we passed, they had sixty alive. Those of Lindeberg's men who were not needed to watch the reindeer had been assigned to pull reindeer sleighs up the route. Each man had three or four sleighs at the end of a rope, and thus they pulled them along by the river. We then went on and arrived at the mission in the evening, while a few stopped over at an Indian hut about three miles from the mission, due to fatigue.

## April 24

— Sunday. Some of us went back with supplies to the men who had stopped over at the Indian hut, while the others remained in the barn to pass their Sunday there. We were not able to drag the reindeer sleighs up the river, due to storms and rain. There were, then, several of us who spent their Sunday in the hut, which had only a few rotten pieces of wood for a floor. Nor was there much in the way of roof or walls. Inside the hut were large puddles with all sorts of filth, old Indian shoes, and bits of clothing, rotting pieces of salmon, feathers, and birds' bones, and broken, old rifles and worn-out chests full of human bones, which made a repulsive odor. But we could not stay outside, due to all the rain and storm; so we were forced to seek some sort of shelter, especially from the cold wind. There was nothing preventing us from coming inside.

## April 25

— We began to drag the reindeer sleighs upwards along the river, each man pulling three sleighs. We could not take along our bedding, but had to sleep on the bare ground under the open skies, so there was little sleeping that night.

## April 26

— We left early in the morning on our ascent, dragging the reindeer sleighs. Late in the evening we reached the herd and our people just above Klukwan, where they had stopped for two days, since there was available a sort of swamp hay that the reindeer ate.

## April 27

— We started off to travel farther, but the herd was left behind with a few men on watch; the rest dragged the sleighs and supplies along the ice. The farther upriver we came, the more ice and snow there were on the rivers. When the snow was hard, we could move up rather quickly, although each man had a hundred pounds in the heavy, flat-bottomed sleighs. We covered about ten miles and reached Redmyer's camp, where we had brought the first supplies when we had all come up on the first trip above Klukwan with the Indian canoes. Here we stopped for the night. Kjellmann was here with us and pulled his own sleigh with one hundred pounds like the others, and ate the same food as all of us.

## April 28

— At four o'clock in the morning, while the snow was still frozen hard, we had to start out and make use of the morningtime before the sun came up and the snow began to melt. Then we could rest while the day was warm.

## April 29

— Since it had rained during the night, the result of it was that we should have a heavy go of it, and that is how it was; and we had to trade places with one another at times as we walked, to relieve one another, for the snow was soft. At dinner time we came to the end of the valley. Here the route goes up a high mountain slope, to reach the top. All who were going to the Klondike went this way across the mountains. We hauled sleighs and loads up to where we stopped at the slope, until the next day. All were tired and fatigued, so there was very little activity, although we were a whole army of people. After we had coffee, Johannes Rauna went into the mountains. He came down an hour later and told us that he had discovered moss on the mountain. There were large quantities of it, which we would investigate further in the morning. We then slept, making two large groups, each under its own large fir [spruce] tree since we had no tents.

## April 30

— We climbed up into the mountains, each with his own sleigh, and found it just as Rauna had said. Each of us filled his own sleigh with moss and then sat on the sleigh. At breakneck speed we went back down to the fir [spruce] trees, and then farther down, all the way to the river. There we emptied the sleighs to go back up for more in the morning.

## May 1

— Sunday. Kjellmann sent two people down to take a message to the men who were with the reindeer herd, that they should try to come carefully up with the herd. Lindeberg, who was the leader at that time and had command, began to guide the herd up after he had received orders from Kjellmann. Redmyer and his men, whom we had passed three days previously, continued to haul up supplies as before. They had been busy at this since the first time we were up with the canoes to Klukwan. Then we others went again to the mountains with our sleighs to fetch more moss.

## May 2

— We had now brought down so much moss that it was presumed that there would be enough. So we filled our sleighs with moss and headed down towards the herd, in order to have some sustenance for the animals when they arrived. We dragged the moss down about seven miles, where we met Hedley Redmyer with his people, who had camped by the river on their way up. Here we rested for the night.

## May 3

— Early in the morning we took our sleighs, loaded them with goods from Redmyer, about a hundred pounds for each man. Thereupon we went up the valley, the same way we had come down yesterday, twenty-six men altogether. Since it is only to be a quick, one-day's journey, there were none of us who took their sleeping gear along; the loads were heavy enough to drag. At night we had to empty the sleighs and draw them over us, although that did not help very much against the cold. We had spruce boughs under us, and as protection from the wind at the sides we put birch bark. But all we have over us are the sleighs, so one can imagine how much sleeping there will be on the snow and ice, where it rained so that everything is wet and damp. Food is not much better than anything else, as one drank the black coffee and ate a piece of meat roasted on a stick; there was not enough bread for everyone at breakfast time next morning. (We had to hold out until supper time, when the cook fixed a tin pail of grain soup with a piece of meat for each man; we would eat it and then prepare for the night.) Sometimes there is grumbling to Kjellmann, although this was of little effect as he was under the same conditions.

## May 4

— We were all up early in the morning, although we knew that it would not help to get up; we have to carry on with empty stomachs until we reach the mountains once again and unload our sleighs, before we could have our breakfast. And that was far into the afternoon. (We should not have had any food then, either, if we had been obliged to stop over there until the next day, when the herd was to arrive.) Late in the day, at dusk, we saw a messenger come; we all were happy, for we learned that we should have a day off after the whole force arrived near where the moss is. But right in the middle of our joy we fell into despair; for just beside the place where we had camped during the night was so much snow that they (deer and men) could barely get through. Since the reindeer can move about easily, the whole herd turned away from the men who had crossed the river with the animals, and all of the deer went into the forest, where it was impossible to get them out again easily. Darkness fell, and Kjellmann again pondered the situation. That did little to soothe us; we could expect a warm day tomorrow until all the deer are found and brought back to the camp. Most of us kept a fire going all night, and sat by it, getting little sleep.

## May 5

— Kjellmann sent some men down with moss for those reindeer that had become exhausted and were left behind on the ice, where there was nothing for them to eat but snow. All others are to go into the forest to search for the reindeer that had scattered in the trees the night before, and to gather them up. I went with the men who head down with moss for the deer left on the river ice. We had our breakfast of black coffee and departed at once, each with his own load of moss. As we encountered the reindeer, we gave each of them moss, and we went on in that manner until we had used up our loads. We rested then, while the sun was warm, and fixed some soup of cornmeal and water. We decided that some of us would return on our route, and try to bring along all of the reindeer that were still alive, of those we had just passed by. The rest of us would go farther downslope, to try to find others that were alive and could be brought up. Rauna and I took the trail upwards, and we began to lead those reindeer we had there where we had just rested, but they had not gone many steps before they lay down and would not walk. We shook them back onto their feet again, but they would go a few steps and then fall down again. We struggled with them until they fell, and died. The weakest one fell first, and then one after another would fall, until we did not have a single animal left out of the group, although we tried to handle them as gently as possible. That did not help, for when we had gone a distance with them, they simply became tired. Then we encouraged them to go farther and they would walk until they were seized by a sort of cramp, which caused the entire body to shake—and within a few minutes such animals were dead. We kept at it all afternoon, with other reindeer along the way, and we tried to reach camp with them before dark, but we came only half the distance to camp and there Rauna and I had to stop and gather wood. We kept a fire going all night, sitting by it and warming ourselves. We had coffee with us but no kettle; but we found a small tin can and made coffee in that. We had only enough for this one time.

## May 6

— Prayer Day. Morning broke forth with sun and clear weather, a lovely Prayer Day morning. We had nothing for sustenance so we fasted. We began to drag and push the half-dead reindeer. As soon as we had begun to move with one of them and had just about reached the next one, which stood and waited, the first deer fell dead. We try to encourage those remaining alive. But of all we have attempted to get on their feet, we had only four alive. I felt certain that about forty reindeer lay dead, just there where we were. When we reached the upper camp, we learned that the men sent to search the forest had been lucky enough to find the herd. They had gone up the mountain with the deer, to the moss place, one mile's distance from the river. Kjellmann had counted the number still alive when they reach the moss; the entire herd numbered one hundred eighty-five, healthy and weak together. The others of the five hundred twenty-one were dead and fallen away. Kjellmann ordered Hedley Redmyer with a crew of fifteen men to start then over the mountains with this remnant of the reindeer, to Circle City by the Yukon River. The rest of the company is to go back, down to the sea.

As soon as we came down from the mountain, all began to hurry along, in order to get back to the mission as quickly as we could, where we could get food. Hans Samuelsen from Lakselv and I ran across the river, where some tents were standing, and where we thought we might buy a piece of bread. I had no money, but my comrade did. But we did not have to buy anything, and instead, we were each given some small rolls to eat. We set our course after the others, as happy as if we had got a whole horse. We still had about fifty or sixty miles to go to find the mission. Each of us was carrying his own bedding, which was not so light a load on a tired and hungry man's shoulders. We walked downwards as fast as we were able, always trying to estimate the shortest route in order not to be left behind. At times we crossed streams so deep that we waded in water that was over the tops of our rubber boots, until we were finally as wet and soaked as if we had been in water over our heads. Eventually, we became so spread out over the route that only one and two were together, and others were left so far behind that it was not possible to see those who were last. Some decided to build themselves rafts to descend the river, and one group had a little boat on which they intended to sail down. Wilhelm Basi, Otto Greiner and I were traveling together on foot, and we were the first ones (foot travelers) to arrive at the place where we all have planned to stop for the night, just above Klukwan. Those who had the boat were Hans Samuelsen, Jafet Lindeberg and Samuel Josefsen. We saw them as they flew past us just above the place where we are to stop; so they were first to arrive at the camp area and we were next. But we were so far behind them, that by the time we got there, they had a fire going and were sitting around it, warming themselves. Lindeberg was busy drying his clothes; they told us that they came to a place where the stream was strong and their boat went under in a powerful eddy.

William Kjellmann.
*Photograph published in Sheldon Jackson's report, 1897.*

Lindeberg grabbed hold of a branch and pulled himself ashore, while Samuelsen and Josefsen stayed with the boat and fortunately came to calmer waters, where they got Lindeberg on board and sailed away again to reach the camp, and the tent where they had dry clothes. Others came in later, one and two at a time; they said that Peder Berg and Jeremias Abrahamsen had stopped upstream to build themselves a raft on which to travel the river. And Balto and some other Lapps were going to do the same, to make the descent more easily. Kjellmann and the other Norwegians walked all the way and in the evening came to the camp in the vicinity of our own, and there they stopped. Not a bit of food was to be found at our camp. A number of men said that they themselves had cooked some meat from dead reindeer. At one place along the way, some containers of beans, which had fallen from the sacks of prospectors, were found in the snow, and these were opened up and cooked and eaten. So we collected some money, five dollars altogether, and Lindeberg, after he had put on dry clothing, was to go to a large camp, which was nearby, to buy food, for the five dollars. He went and explained how we were, and for nothing—no payment at all—he obtained some flour and some grain and meat. We cooked these and had our supper. Kjellmann could have bought supplies, but he had done that before and probably was reluctant. So he hungered as we did, and I think that there was not very good supper food in Kjellmann's camp, either.

## May 7

— Before we managed to leave the camp, Peder Berg and Jeremias Abrahamsen came in. They had almost gone under, traveling down the river on the raft that they had made, at the same place where the boat had gone under; they escaped with their lives but lost their traveling packs, wherein they had clothes. They were wet and miserable, and without food, when they came into our camp. They took in a little sustenance, and then Peder Berg decided that he did not want to lose his pack, so he returned back upriver in order to look for it, while Jeremias Abrahamsen went with us. At the same spot on the river, it had gone in the same manner with Samuel Balto. He nearly lost one of his men, but they still continued down on the raft through the night. We left our camp as we had arrived, one after another. Those who were strong in walking arrived first at the mouth of the river. Those were Kjellmann and one Lapp and some Norwegians. In order to get across the river, we waited for the boat, about half an hour. Once across, from there to the mission was only a few miles, which we covered in plenty of time before dark; but a number of men did not come in until evening and late at night.

## May 8

— Today we observed our Sunday peacefully, and supplies for us were weighed out, too, so that we did not have to starve any more; and some of the Lapps ate until they felt sick. We had previously stayed in the pastor's barn, but now we were to live in tents since it was summer and good weather with warm sunshine. Peder Berg came in, carrying his pack; he had found it alongside the river. He had been delayed by his search, and thus he was the last man to arrive.

## May 9

— All of us worked together, gathering all of the sleighs and in building a house where we put all the goods to be left behind here. Those goods that were to be carried along were to be packed together, in bundles, and made easy to handle.

## May 10

— Kjellmann traveled to Dyea, to arrange for the ship to call at Haines and take us on board in order to go to Port Townsend. We get to rest a little now from the hardships of the mountain journey.

## May 11

— All of the soldiers who earlier were stationed on the sandy beach, as well as the officers, were under orders to go to the Klondike, presumably with the reindeer. But when things took a bad turn with the reindeer, none could come to join us, and they were called to the war in Manila. They shipped out in the forenoon with music and hurrahs.

## May 12

— We have nothing to do, just to walk around and observe how the Indians live, and watch when they come ashore from fishing. The two men who have lain sick left the hospital and are well again.

## May 13

— We have good weather with sunshine and nothing to do. Nor were we industrious, as all have such sore feet from the terribly long march in heavy, rubber boots that we can scarcely walk.

## May 14

— We received orders to be ready by eight o'clock tomorrow to board ship, as a steamship from Dyea is going to pick us up on its return trip.

## May 15

— The ship arrived, and we went aboard, where we are treated royally. We even got to eat in the dining room among the Americans. Several hundred men on board came from Dyea and Skagway; they had to return when their money ran too short for them to travel over the mountains to Dawson. In the afternoon we came to Juneau, a lovely little town that was planned and built after gold was found on an island in the vicinity, called Douglas Island. There lies the richest gold mine in the world, which is now owned by some capitalists in England. We departed from Juneau after a few minutes' stopover.

## May 16

— We had conversations with a number of Norwegians who had also come from Skagway, and who told us about various conditions there. In the first place, they said that there was no work or way to earn money there; so those who were on board were basically people who had been there and gambled and had drunk up their money, so they had to go south again. Many murders, robberies and thieveries were committed, all for money. One was robbed of ten thousand dollars, and no one knew who the robbers were, for most often such fellows appear both fine and respectable, in clothing and in speech. A woman was murdered and her money was stolen, four thousand dollars; and no less than five of such fellows who were on board had been there on such business. They appeared older and much like gentlemen. At dinnertime we came to Fort Wrangel, which is a fishing town only, where there is a cannery that purchases the fish from fishermen and cans it in hermetic containers. It is sold at large markets. The fortress is called "fort" in English. The town has the title of "fortress" from the Russians who had their fortress with cannons at Fort Wrangel when Alaska belonged to Russia. We stopped there for a few hours and we had time to run ashore on a quick trip. The ship departed after that.

## May 17

— Norway's national day! We all remember the Seventeenth of May. Many among us said that there should have been a dram on this day; but all are without money, so the one could not help the other. Thus all had to go without the dram to honor the Seventeenth of May.

## May 18

— Our ship called at the town of Victoria, which lies on Vancouver Island and is the capital of British Columbia. It is not large, but it is beautifully situated at a convenient and lovely place. We had three hours to run into town, but three hours were too short a time to see much. We could not run too far, otherwise the ship could go without us. Not far from Victoria lies the little town of Port Townsend in the state of Washington, where all of the families of the Lapps were put ashore when we traveled northward. We are to go ashore there, and we arrived in the evening. So the journey south from Haines Mission took only four days to Port Townsend. We had a fast and enjoyable trip; our ship went seventeen miles an hour and the time passed quickly and pleasantly.

Kjellmann found a little steamboat that led us across the harbor to the other side where the Lapp families live. We arrived late in the evening, after everyone had gone to bed. The ship anchored at the long wharf that extended out into the sea, and blew its whistle a second and a third time, whereupon someone at last came walking down with a lantern in hand. We could have walked, but it was so dark and all were unfamiliar with the place, so we waited until the lantern came. It was carried down to us by the foreman Regnor Dahl, a Norwegian who had got that position because he could understand a little Lappish. He showed us to a house where we might sleep, and some of the women fixed food for us.

## May 19

— First of all in the morning, we had to see which edge of town we had come to, and how it looked. We found this place to be excellent. It had been a station for soldiers for some years. The fortress was built by the sound, on a high sandhill right across from Port Townsend, and it is still under construction. We had a good view of it in all directions, and in addition we could see the large, level meadow that was for a recreational area. If one wished to go to town, it was only a one-hour's walk. But Regnor Dahl had built a boat, pointed at the one end and square at the other, about eighteen feet long and at the widest ten feet; so always there was a means, in case anyone wanted to go by sea; except that one had to count in, that although the distance was shorter by boat, it nonetheless took more time to get to town by that way than going on foot around the shore. At this time, Kjellmann gave all authority to Regnor Dahl, who was to command and keep everything under control, and provide food and clothing for us. If any Lapps should become drunk, Magnus Kjeldsberg and Jafet Lindeberg had police authority to place those fellows into the jail, which stood only a few yards from the houses where we lived. Kjellmann then left us, as he was going to travel to his home, to his wife and children in Wisconsin. As soon as navigation to Alaska is open, he will come back. He said, however, that it could happen that we should not get to see him before we came to St. Michael in northern Alaska. Thereupon he left.

## May 20

— Everything was quiet, and most of our work was only to walk about and enjoy the peacefulness, although we have had some things to do. The sun was too hot for us to work, for we are not used to that heat, so thus we saw gladly that we had time on our hands; nor was our pay any greater or less.

## May 21

— A trip to town by one of our Norwegian comrades, as he himself told it: He went to town yesterday, as he had such good opportunity to get a ride with a waggoner/drayman who came from town with our things and brought them from the steamship freight and luggage office, here to our place. He then rode with this drayman to town and planned to return for supper; but when he came to town, he met someone with whom he became acquainted and they joined company to look around in town. Among other things, they went to the Indian village on the edge of the town, in order to see the place. His companion, without warning, aimed his revolver at an Indian boy and fired, whereupon the Indian boy fell with a terrible cry, and the shooter's companion stood as if the bullet had struck himself. The shot had gone through both of the boy's thighs; and the doctor was fetched immediately. No police or other people were present, so there was no investigation or questioning of the criminal—the man who did the shooting. It became too late in the evening to return home, so our comrade spent the night with that same man until the next day; but he did not learn the reason for the deed. Thus he told of his trip to town just as soon as he returned in the morning.

## May 22

— Sunday is a lovely day with sunshine, calm and clear weather. When one sat or stood still and listened, one could hear snakes hurrying through the grass, but they were not large. The largest I happened to see was about twelve inches long. We kept ourselves in the shade all day, as the sun was too hot; we were not accustomed to such.

## May 23

— We have just the same life and monotonous conditions as the day before, with sun and heat, which is not remarkable, since Port Townsend lies at about the latitude of Paris, where the sun shines passably warm at this time of year.

## May 24

— Since we have had it lonely and quiet for a few days now, it was time to have a bit of a change again; and thus it happened, too. Some Lapps had got whiskey from town, and in the evening they became intoxicated and created a scene and disturbance in others' rooms. A complaint was lodged with the foreman, and one of the Lapps by the name of Johan Rist was put in jail, and would have a hearing the next day in the forenoon, to investigate the matter further as to whether he was the one who had created a disturbance in Jakob Haetta's room.

## May 25

— Hearing and court action. There is to be a hearing for Rist, whom Dahl had thrown in the jail in order to investigate more closely his case for acquittal. The room where the Norwegians lived was selected for the courtroom as it was the largest. Everyone was to have free admission, women as well as men, and even children. The room was furnished with benches and a long table in the middle of the floor. At the table sat Regnor Dahl as counsel, and Brown, an American, beside him, who was there to look after the place, while jurors Ole Berg, Lauritz Larsen, Ole Krogh and Per Porsanger sat on another bench beside them. At that point the judge advocate gave the order to the police that they were to fetch Johan Rist out of the jail and to lead him before the court. A moment later the police came into the courtroom with the arrested person, and he was shown to a bench where he was permitted to sit. Then the judge advocate read for him the charges of which he was accused: having conducted himself brutishly and improperly in Jakob Haetta's room, as well as breaking windows in another room. After that the first witness was summoned, who explained that Johan Rist was intoxicated, and that he stood not far away when Rist broke the window in one of the houses, and that when Rist knocked on the door of Johan Tornensis' room in order to go in there, he did not get in. The next witness, as best I observed, was Inge Bals; she was called in as it was presumed that she was said to have the most reliable testimony regarding Rist's behavior the previous evening. When she had spoken, it did not agree with the first witness, as it also could be proven that she said something at that time that was not true. Next came the third witness, whose testimony did not completely agree; and the fourth witness was dismissed by the court, being presumed to be sympathetic to Johan Rist. The judge advocate had written down what each witness had said and testified against Rist, and read it aloud in Finnish for Rist. At this point the judge advocate enumerated some points from the law of the United States, regarding domestic peace, which went as follows: "The American law is such that, when a man or person is in or resides in the house, that same person has the power and authority to keep the law and order in the same house, whether the same person owns or has rented it;" (so I, for my part, find Johan Rist to be guilty under this statute, and to be liable for punishment, for breach of the peace against Jakob Haetta and Johan Tornensis) and "The jurors must pass judgement in the case, wherefore all observers must leave the room." And Johan Rist received no defense, nor was he allowed to defend his case himself. The jurors decided in the case, and the people were called back in, whereupon the judge advocate read aloud the sentence, which consisted of two days' bread and water. The judge advocate ordered the police to take Johan Rist to the jail and put him in. Afterwards, all of the people followed while Rist was led to the jail to suffer for his crimes. And his wife, along with other Lapp women, also went along, and cried. Rist was put in the jail, according to the sentence of two days' bread and water. But as evening drew on, Regnor Dahl had no peace from Rist's wife, who begged with tears for mercy for Rist, who sat in the dark cell without food. So Dahl had to go immediately to discuss the matter with the jurors, whereupon they of course gave their consent at once, and word was sent to Rist's wife, who was sitting and crying for the sake of her husband, and she was informed that he was to be set free. Regnor Dahl, the policemen and a number of other men went to get him out of the jail. As they returned, Rist was walking right in their midst, and a crowd of Lapp women followed, so that it appeared quite solemn. Regnor Dahl turned Rist over to his wife, who led him home to their house.

## May 26

— Everything was peaceful and quiet. All we did to pass the time was to play ball on the meadow, or now and then to take a trip to town.

## May 27

— Nothing special happened today, either. It was as if everyone was depressed or had become a bit withdrawn after Johan Rist was arrested.

## May 28

— The houses were cleaned, inside and out, in order to have them nice and clean for the Pentecost holiday.

## May 29

— Sunday, Pentecost. Some traveled to town and others took a trip into the country to see how the farmers lived. I went along on the trip to the country. We took a round trip of several miles across the countryside and found everything beautiful and lovely.

## May 30

— Some of the boys went to town and brought us letters from Norway, with many pieces of information from home, which was not forgotten by anyone.

## May 31

— Nothing noteworthy; everything well as usual.

## June 1 to June 5, Sunday

— Everything was quiet, so it began to seem quite dead at the station. The hot sun made people so sluggish and lazy that we felt best when we lie down. Regnor Dahl received a telegram from Dr. Sheldon Jackson, saying that he would visit us tomorrow.

## June 6

— Everyone, all day, waited for Dr. Jackson. All the Lapps dressed themselves in their holiday costumes, and Regnor Dahl went about dressed up a bit, himself, but no Jackson came this day.

## June 7

— Since Dr. Jackson did not come yesterday, we think that he will certainly come today; and while we waited all day, still no Jackson came.

## June 8

— A little, white steamboat came to the pier, whereupon Regnor Dahl strolled down to receive guests. As soon as the first man walked up onto the pier, we immediately recognized Dr. Jackson. With him was a person whom we did not know, but later we learned that he was a lieutenant (military). Those three then came walking up—Dr. Jackson, the lieutenant and Regnor Dahl, and they went into Dahl's house. Later came a message from Dr. Jackson that all of us should meet at Regnor Dahl's house, as he had something to say to us. We all hurriedly gathered there in a group, in order to hear what our esteemed leader would announce. When we had assembled together outside of Dahl's house, Dr. Jackson came out, and bowed to us. Thereupon, came also Regnor Dahl, who was going to translate into Norwegian what Dr. Jackson said in English. This was: Dr. Jackson has hired a large sailing ship, which shall take us to St. Michael, and which was now at a wharf in Seattle, being checked to see that the ship is seaworthy. Dr. Jackson said that the ship was strong, for it was only three years old. It would be ready for departure on the nineteenth of June from Port Townsend. We asked Dr. Jackson then why we could not go by steamboat. To that, he answered that we should never have it so good on the steamboats as we should have it on sailing ships; and further, that the sailing ship would be in St. Michael before all the steamboats. We did not believe that, since we had already learned a bit about Dr. Jackson's good promises, that had not been carried out or upheld. Dr. Jackson certainly noticed that we had lost faith in him; and when we opposed his decision to go to St. Michael by sailing ship, which he without a doubt had thought we would do, he had to offer two men free passage to Seattle in order to look at the ship and see if it was good and sound. When our meeting was over, Dr. Jackson visited one of the Lapp families, which according to the doctor's report, had a contagious disease and should not be sent to Alaska. Dr. Jackson gave the head of that family permission to go back to Norway. He then gave orders to Regnor Dahl to move the Lapp family to town in the morning, in order to go by steamship to Seattle, where the head of the family would receive a settlement of his contract, signed in Norway, and receive pay from the contract day until the last of June. He would also receive pay for everything that he had sold in Norway to the United States government in the way of reindeer, reindeer gear, sleighs, fur articles, etc., as well as free passage home. The Lapp family had a grown son who would remain behind, hired by Dr. Jackson on behalf of the U.S. government, to be part of the crew going to Alaska.

## June 9

— Dahl ordered some men to accompany the Lapp family to Seattle; the family is to go back to Norway by way of Port Townsend and from there by steamship to Seattle. At the same time, Magnus Kjeldsberg and Jafet Lindeberg also went to Seattle in order to inspect the ship that was to take us to St. Michael, as Dr. Jackson had said that they should determine whether the ship was substantial and whether there was room for all.

## June 10 to June 12, Sunday

— Everything was peaceful, quiet and orderly. Magnus Kjeldsberg and Jafet Lindeberg came back from Seattle, where they had seen the ship on which we were to travel to Alaska. The ship was strong and new, but too small to house all of the people and the supplies that should be taken along. Now, if one wishes, one could set oneself against the journey and not board the ship when it arrived. Plans were made for the best way of traveling north by steamship to St. Michael.

## June 13

— It was quiet and peaceful at the station. Except that a few Lapps, who still had not drunk up the money they had with them from Norway, went and drank improvidently, for they should soon leave "whiskey" and begin on the tour to northern Alaska where spirits are not so abundant as in the United States.

## June 14

— Regnor Dahl's power and authority were great, which led to more life and amusement in our camp. Nearly all of the boys were in town and did not return before evening. Some were quite "juiced up" and that delayed the homeward journey until darkness had fallen. Jafet Lindeberg had also had a drop to drink. Regnor Dahl wanted to arrest him; but Regnor Dahl himself was drunk, and in addition, he had only one arm. So it was impossible for him to do that alone. Dahl stormed out; he could barely walk so he got himself some help against Lindeberg. I shall leave it unspoken as to whether it was hired or voluntary aid, but Regnor Dahl finally got Lindeberg in jail. Regnor Dahl should have been put in jail long, long ago—he only did this to Lindeberg out of malice. (While I now speak of Regnor Dahl, I shall also make clear that he was not only the foreman but also supply master, who handled many matters in less than a just manner. He also threatened many times to feed us in the same manner as we had been in Alaska, at Haines on our mountain journey; so Regnor Dahl was little respected by the people.) Now Dahl thought that everyone could quietly go to bed for the evening, but things were not yet over. A Lapp, Anders Biti, had got himself drunk. When in this drunken state he had struck his wife and when this unruly man even had shown a knife against his own wife, it was reported to Regnor Dahl. Dahl had recommended that Biti be arrested, and he was. But when they were going to place him in the jail, he broke loose and ran down a high and steep clay hill, just down from the detention house. The hill was forested, and no one could see where or which way Biti fled; it was also dark. They watched and waited a bit, to see if he would come out so that they could get hold of him. Finally they, for the second time, were able to seize him—he was with his wife, up in his own room, where he had sneaked unnoticed, through forest and darkness. Thereupon they took him and led him once again to the jail and put him in, where he sat all night as punishment for his roughness.

## June 15

— The arrested parties were set free, but no court action was set in motion. Jafet Lindeberg wanted to have his case before the court; but Regnor Dahl feared that many would support Lindeberg; so the court was suspended. Anders Biti had nothing to say, since he acknowledged his guilt.

## June 16

— We had some work at packing some goods together—furs, etc.—that were to be taken to Port Townsend as soon as possible. We received news that Dr. Jackson would visit us tomorrow, to concern himself with our group, as Dahl informed us.

## June 17

— Dr. Jackson came in the forenoon, as he had announced for his arrival. But this time, he came to fetch those men whom he was going to send to Siberia and to St. Lawrence Island, which lies in the Bering Sea. Jafet Lindeberg, Ole Krogh and Johannes Rauna were ordered sent to Siberia, to the government's station, to remain there for the winter. That Lindeberg was ordered to Siberia was certainly due to Regnor Dahl, but Lindeberg made no resistance, nonetheless. Rolf Wiig was named for St. Lawrence Island, also due to Regnor Dahl's false reports. These men then traveled with Dr. Jackson to Seattle, to go from there by ships to northern Alaska. Before they departed, we asked Dr. Jackson about citizenship papers, or perhaps to let us have enough money that we could apply for them ourselves, to which he gave his word that he would arrange and pay for them. Regnor Dahl is to show us to the place where we could get our citizenship papers. This is to take place in the morning, when the whole group of us is to go to town. We have Magnus Kjeldsberg to thank for having arranged all of this for us. He was always insistent that we must not leave for Alaska without our citizenship papers, for without them we should have no right to government land, but when we have our papers, we can claim land anywhere, where no others have the ground, and retain it as our own property.

## June 18

— We marched into Port Townsend, a whole crowd of young men, to take the oath of induction into citizenship and to swear our allegiance, from King Oscar the Second, to the government of the United States. The oath was carried out as all raised the right hand, all at the same time. Thereupon a man read an article from a book; and none of us understood what he read—it was only a short piece. At this point we all put our right hands down, and therewith the ceremony was over. Each of us received an oath paper or certificate that we were now citizens of the United States.

## June 19

— Sunday. This was a rather unpleasant Sunday. In the first place, the ship was supposed to come and pick us up and did not come, about which some were not in the best humor. In the second place, Regnor Dahl let us go without food all of the long Sunday holiday. So we felt irritated by more than that the ship failed to pick us up. We seldom grumbled about troubles we only imagined.

## June 20

— Since there was no visible sign of the ship, Regnor Dahl had to go to town and buy a little food for us, otherwise we should have felt starved to death. When he came home from town, he explained that Dr. Jackson had stopped the delivery of supplies for the reindeer expedition, so he had to telegraph Dr. Jackson in order to have some things delivered, which he finally managed to do. But it seemed that all of this is to placate us and to fool us about his dealings in the administration of supplies. This evening, the ship is to be at the pier to take us on. But that was not fulfilled, either, whether it was a telegraphed fabrication by Dr. Jackson, as Regnor Dahl told us, or a falsity by Regnor Dahl; and we got no more food than just sufficient for the day.

## June 21

— The ship came at last, a little schooner the size of an ordinary Norwegian fishing cutter. We are to be stuffed in there, about ninety people. At first Regnor Dahl agreed with our contention that it would not do for so many people to be crowded into such little room, which was already crammed full of goods. We found, in addition, that thirty-seven men were already on board from Seattle, after Dr. Jackson had told us that no other passengers would be permitted along on the ship. Of course, it may be possible that they had been allowed to board without Dr. Jackson's knowledge. When we therefore were reluctant to board, the captain had no recourse than to try for the support of Regnor Dahl; not long after they had talked in the cabin for a while, Dahl's voice took on another tone as he tried his best to get us all on board. It ended up that there were only five Norwegian boys—Thoralf Kjeldsberg, Ole Rapp, Lauritz Stefansen, Peder Berg and Jeremias Abrahamsen—who did not go aboard. Instead they walked into town to look around; the ship would soon sail, but they were indifferent to that. Rather than going with the schooner, they all agreed to hire themselves out as soldiers in the war, in order to go to Manila, or to go to sea. They had no more than one dollar, as they themselves said; but then Thoralf Kjeldsberg was good in the English language, so they could speak with people. The ship set its sails and made for Port Townsend, where it was to stay over night. At ten o'clock in the forenoon tomorrow, the ship is to sail regardless of whether the other boys came along or not. So it seemed a bit strange, and as if several others were on the verge of being lured ashore in order not to be separated from their companions. But we decided to let it be until morning, after we spent the first night aboard the *Kenney* [the *Louise J. Kenney*].

## June 22

— Morning came with good weather and a good breeze for Alaska, so we crept out of our cramped sleeping compartment early and went to the deck to see what would happen. We saw the captain strolling on the deck, so we could suspect that he was concerned about something, but we could not guess what it might be. We presumed that he thought about the boys who were not on board, and how far he could go without having them along or reporting it to Dr. Jackson. At ten o'clock, the *Kenney* still lay quiet, and shortly afterwards came word that the captain could not weigh anchor for a while, although we did not know for how long. Now we imagined that the boys had been active while on land; half an hour later, a little, white government steamboat left shore and headed towards the *Kenney*, and we thought that something was afoot and it could be entertaining. The little steamboat came alongside the *Kenney* and three men came aboard our ship: the customs chief, the consul and the health commissioner. These three went around the ship and measured it, which was done in a few minutes. The ship was then impounded, and later, Dr. Jackson arrived. We thought that the boys ashore had accomplished this, and also caused a good article to be placed in the American newspapers about the treatment of some helpless people, for the possible advantage of the interests of the gentlemen who could not understand up from down. Dr. Jackson had read the piece in the newspaper's columns in Seattle just before he received a telegram from the consul, asking that he come to Port Townsend and take care of his business regarding the Laplanders. We thought that Regnor Dahl saw that all this was clearly going to mean other arrangements were to be made, and that he knew that all of us realized that he had allowed himself to be influenced by the captain in order to get us on board yesterday. Now, in order to get on the good side of the people, he said that it was he who had telegraphed Dr. Jackson to come and take care of the irregularities, inasmuch as the people would not go on board. It was certainly not the first time we had heard such excuses. Dr. Jackson came at two o'clock in the afternoon on the ship from Seattle and with him came Wilhelm Kjellmann. They boarded the *Kenney* and went immediately down into the cabin with the captain. After a few minutes, Wilhelm Kjellmann came on deck with a placard in his hand, from which he read the names of all of those who were to go ashore. About half of the people were removed, and there was even so no more room than was needed for those who remained on board. Wilhelm Kjellmann looked over the supplies as well. It was found that adequate food was not aboard for those who remained on the ship, let alone for the entire company. Kjellmann then ordered aboard the supplies that were needed, after which the ship was ready to sail. All of the Norwegians were set ashore as well as some of the Lapps, and there were about thirty of us who were to go to Seattle. We departed from Port Townsend after four o'clock in the afternoon. During our trip to Seattle, Kjellmann told us about Hedley Redmyer's travel with the reindeer; he had had to let eight men return from the mountains, for they did not have sufficient supplies for all the company to cross the mountains to Circle City.

Olai Paulsen and seven others had to leave, to return to Haines. They had a three days' journey, endured without the least bit of food. The rivers were swollen then, for it was just at the time of spring runoff. Lacking supplies, they could do nothing but make a raft so as to go more quickly downriver, whereby they were nearly killed. But they made it down to Klukwan, and there Paulsen had to buy a small Indian boat in order to get to Haines Mission, where they could get food and not starve to death. They came then with good fortune, to the mission, from where they went by steamship to Seattle. In Seattle, they met Jafet Lindeberg and thus received news of how everything stood. Later they contacted Dr. Jackson, who took them under his care. Olai Paulsen was given leave to go home to Denver, and to come back later to Alaska. Paulsen reported that one hundred sixty reindeer were alive, and were healthy and fat.

## June 23

— Early in the morning we arrived in Seattle, where we received our quarters on board a little wooden steamship [steam schooner] called the *Navarro*. This ship was to take us to Alaska. We saw that the *Navarro* was no better than the *Kenney*, but now we had to hold our tongues, since we knew that there would certainly be no change. We especially wanted to keep quiet when we heard that a number of Americans were also going to go north on the same steamship. About our own group, we learned that after Dr. Jackson had ordered Rolf Wiig to go with a missionary priest to spend the winter on St. Lawrence Island amongst the Eskimos and to hold school for them, Wilhelm Kjellmann exchanged Rolf Wiig with Jeremias Abrahamsen; but he let the others go as Dr. Jackson, and Regnor Dahl, had ordered.

## June 24

— We spent the time in town by looking around on the streets and in large stores. There was recourse for those who had only a few cents; some of the Lapps had sold some winter clothes for next to nothing, only to get whiskey. Many were gotten drunk "gratis," just so that people could have some amusement from them. What sort of people are these, who always try to get others intoxicated? It seems they look for foreigners who come into town and then many of them try to lead newcomers into drunkenness and into out-of-the-way places in order to be robbed. We learned that others try to get the men into their houses, while they are drunk, and there they have a sort of anaethetizing drop that they mix with whiskey and give that to a man, to drink, after which the victim falls asleep and is taken aboard a sailing ship that needs to fill out a crew. Those unfortunates do not awaken until they are far out to sea. One of the men from our expedition was subjected to such fellows, who got him so far that they locked him into an out-of-the-way room. When he realized that he was locked in, he opened a window facing a street and saw that he was on the second floor. As he was considering what should be done, he heard people coming to his room, whereupon he crept out of the window. And just as he had crept out, three men came after him. He leapt down into the street. Lucky and well, he ran off without being injured, after which he came aboard and told us of this event. It happens always in large seaports that many men and young boys are robbed in that manner, and even murdered. The fault of everything is that terrible whiskey, by means of which are suffered great destruction and misery; to it even older and experienced men are drawn, although they conduct themselves with great caution.

## June 25

— The steamship *Del Norte*, which was loaded with supplies for St. Michael, Unalakleet and Golovnin Bay, as well as Siberia, departed after dinner [in the afternoon]. A group of people on board was going up to Alaska to look for gold, among other things. Jafet Lindeberg, Ole Krogh and Johannes Rauna, who are going to Siberia, as well as the priest and Jeremias Abrahamsen, who are going to St. Lawrence Island, went together on this ship.

## June 26

— Sunday, St. John's Day. We had said constantly all spring that we had to arrange things so that we could see how it would be to spend a holiday, such as St. John's Day, in America. At least we can observe the holiday quietly, but in Seattle there is no difference between St. John's Day and Saturday.

## June 27

— According to the plan, today was to be the day of departure for the *Navarro*, but stores had to be loaded on, so we do not leave on this day. Dr. Jackson and Wilhelm Kjellmann were not going to go themselves until after we had departed; they were to travel on one of the best of the ships going to Alaska.

## June 28

— At dinnertime the *Navarro's* whistle blew for the first time, and everyone had to go on board. It was strange to observe how the Americans took their leave of their families, friends and acquaintances—how they kissed each other, both men and women. Then the *Navarro* whistled for the second time, and immediately after that for the third time, whereupon the ship glided out by the dock, while the entire quay was crowded with people who waved farewell with their pocket handkerchiefs. A few minutes later, our ship was going full speed across the sound, so that we were passing our home at Port Townsend already this evening.

## June 29

— We headed to sea through the strait of Juan de Fuca, by Cape Flattery, and there a fresh wind blew against us, buffeting our ship to all sides, such that it seemed that it would bounce to pieces.

## June 30

— During the night our ship had to turn back, because the riverboat that we were towing, became damaged (a leak). When the people were being taken off the boat, one of them broke his foot, and also the captain of the riverboat (the *Minneapolis*) injured one of his hands (a slight injury). Because of the darkness, we anchored about seventy miles from Port Townsend.

## July 1

— Repair of the *Minneapolis* was considered, but in the end we were compelled to return to Port Townsend; there the damage was repaired and everything put in order, once more to attempt to put out to sea—and to find out if the boat would go to pieces.

## July 2

— The captain was reluctant to put out to sea from Cape Flattery, and so he telegraphed for a pilot from Seattle, and the pilot came as fast as possible, to take us up the inner passage to Juneau; and we should not go out into the open sea until after that time. We were quite satisfied to learn this; although traveling in the inner passage would take several days longer, we did not care about that. For to set straight out to open sea in a small wooden sailing ship that was overloaded with supplies and had a lot of people on board, and to head three thousand miles across the sea, when our ship made only five miles [an hour] in calm, good weather—what would happen in stormy weather?

## July 3

— Sunday. All were quite happy that we are going to go, from Port Townsend, north up the inner passage instead of heading straight out to sea, to battle the waves and storms. We now know this way, for we had traveled exactly the same route on our round trip to Haines Mission during the past months. Today, an American missionary gave a speech out on deck, after which some songs were sung by three American women, who were also going to Alaska. Of course, we did not understand many words, either of the speech or the songs, but there were about fifty Americans on board, and they understood the words.

## July 4

— The Fourth of July is America's greatest national day, perhaps even the most solemn of all the days that are celebrated, and many sorts of festivities take place. The lay preacher who gave a talk, from the Bible, yesterday, today gave a speech to us, based on articles in an American newspaper. All clapped hands, and the people sang their national songs, after the preacher had finished.

## July 5 to July 8

— Everyone was well disposed, and pleased, in that we were going along through the narrow waterways, between high and low forested hills without feeling the sea or the waves; and that there was good weather with warm sunshine every single day.

## July 9

— In the morning we came to the town of Juneau, where all the passengers were permitted to take a trip into the town while the ship lay in the harbor for a few hours, to take on water and to get items required on board. Then, the pilot went ashore, and our ship headed into Cross Sound in order to set out to sea from Cape Spencer. But when we had come halfway across Cross Sound, we found thick fog, and the ship had to anchor asea, for the night.

## July 10

— The *Navarro* departed early; for when we awoke we were nearly out of the sound. We saw the wide Pacific opening up, the farther we came towards it. After a few hours we were out in the open ocean, so far that in the evening we could no longer see land.

## July 11–July 13

— We were lucky enough to get a good wind. And since the *Navarro* had set all her sails, she made very good speed. We could not have asked for better weather for traveling than we have had these three days, and it looked as though that would continue as we sailed onward.

## July 14

— The wind still blew strongly. We saw land, some islands that lie off Kodiak Island some distance in the sea; but we could not see clearly because of fog, and the islands disappeared from our sight.

## July 15

— Finally we have a head wind and fog, but we now know that we are not far from land. I think that there could not be heavy seas, for the sea and the wind fell from the land.

The steam schooner *Navarro*. More than thirty people of the expedition spent the month of July 1898 aboard the ship en route from Seattle to St. Michael.

*Photograph courtesy of the Puget Sound Maritime Historical Society, the Museum of History and Industry, Seattle.*

## July 16

— We had more wind than yesterday, so the poor little *Navarro* groaned such that the crew had to set supporting beams under the deck in order to support it a little; otherwise, it seemed that the whole deck would have gone overboard. But it held, though just barely. Had the seas been any heavier, I had no doubt but that the *Navarro* would have lost the entire main deck. Still, our luck was good, as usual, and everything went well.

## July 17

— Sunday. Sunshine and good weather, and when we came on deck in the morning, we saw land not far away. At the same time we saw a volcano, a vent at the top of a mountain, from where smoke and steam rose, which we watched with much attention. The closer we came to land, the clearer and more interesting it became to observe. The land, or (rather) the islands here, appeared to be a chain of high mountains, hills and valleys, as far as one can see, with areas of forest of some sort of trees.

## July 18

— We came to Akutan Pass, which leads into Dutch Harbor, where we arrived about midday. We have the opportunity to observe the character of nature here, which is unusually lovely, with high, mountainous slopes covered with grass and without trees. The hills are clad in green from the shore to the highest peaks, so that one cannot see any rock at all, except on the headlands facing out to sea. In my opinion, Dutch Harbor is one of the most beautiful places we have been, along Alaska's coast. At the foot of the fjord, at the harbor, is a lovely little town, inhabited mostly by "mixed folk"—a mixture of Eskimos, Russians, and other white people from the United States. A Russian mission is here, and an American one, as well as some large companies, which have established stores, stocks of coal and water works, where all ships can get water, coal and other supplies. Our ship took on water, so we have several hours' shore leave, and we ran all about and looked around. When we came back aboard, we all were carrying bouquets of flowers, and were enjoying their fragrance.

## July 19

— Early in the morning we departed from Dutch Harbor, and it was only a few hours before our ship was back at sea, to leave the land behind and not to see it for a week, when we shall come to St. Michael. We began on our sea journey across the Bering Sea with a favorable wind and good weather.

## July 20

— Good weather and good wind with rain.

## July 21

— We had calm weather and sunshine.

## July 22

— We had calm weather and fog.

## July 23

— We had calm weather and sunshine.

## July 24

— Sunday. According to what the helmsman told us, we should get to see St. Michael today. But instead of that, we got storm and a head wind, so that our ship had to lay by all day, until late in the evening when the storm abated. The riverboat that we were towing was not strong enough to withstand such a storm and a head wind.

## July 25

— At last we toiled our way into St. Michael in the forenoon. After dinner, we were all set ashore, where Dr. Sheldon Jackson showed us to a little, old outbuilding; therein was neither stove nor anything one could build a fire in to keep warm, and we were all wet from the rain. Nor were there berths to lie on; we shall have to lie on the floor, which consisted of some wood thrown down on the marshy ground; all was wet and miserable. We had not exactly expected to get a hotel in Alaska, but on the other hand, there were many other houses better to put us in than this particular box, which stood out on a promontory on a wet marsh, far from where people lived. We had food for supper; but if we had not asked for it, we should have gone without food for Dr. Jackson's sake until the next evening, and not received any then, either. But since we asked for food, we received some mildewed crackers and roast beef (a sort of canned ox meat). That was all we had for supper. Then we sat there in a cluster, inside of this box; for there was no intention of going outside, as it rained all through the evening until we became sleepy; whereupon we crept into our sleeping bags, thus to make the time pass more quickly and to awaken in the morning to another, more favorable, existence—or, more likely, to the same.

## July 26

— The first thing I decided to do today was to look around in the so-called town, among the Eskimos, whom I had heard about even when I was in Norway. I wanted to see how they lived in their tents and houses, and if things were as I had imagined them. The town consists mainly of two companies—the Alaska Commercial Co. and North America Transportation and Trading Co. Otherwise, there are no stores, only some buildings that are owned by people residing there with their families. Many Eskimos have houses there, in which they live during the winter; but during the summer they live in tents. In addition, along the shore stood a large number of tents, in which were gold miners who there on the sand were camping, in order to go up to the Yukon, and some who had come down the river to go to the United States. After a resolution by Congress, St. Michael has been made the headquarters for the military district encompassing the surrounding places. St. Michael was built by the Russians in years gone by and has only three hundred inhabitants. I had an opportunity to see in what manner the Eskimos ate; I happened to be by their tents, or their summer "camp," right at dinnertime. They sat around in the tents, each with his own piece of dried seal meat, which they dipped in raw seal oil. (I thought at first that it was raw cod-liver oil.) When they had eaten, they took one cup or several cups of tea to drink. This was their dinner. It certainly looked a bit unpleasant to me, and there was a vile odor—not only in the tents, but one could smell it far away. I looked at their boats, which were interesting to see, especially when they sat in them in the water and rowed. The boats are built of sealskin, with fine workmanship on the inside, where the hide is fastened. Each boat is pointed at both ends, tapering from midboat where it is roundish in shape (in cross section), and the average length is from ten to twelve feet; but they are only one and one-half to two feet in greatest width. When they go out in the water with these boats, only one person has a place, right in the middle of the boat, and he sits on the bottom. This kind of boat (which the Eskimos call a "kayak") is covered with hide, except for a round hole in the middle, in the hide covering, wherein the man or woman sits and rows. One can see nothing but the head, arms and the oar blade with which they row, when they are "kayaking." In calm, good weather they can travel several miles an hour, for the boat is constructed specifically to travel easily through the water. The Eskimos differ from the Indians that we saw earlier, in that they are not entirely so dark in skin color; nor have they such sharp features as the Indians, but look rather gentler. Whenever they meet people, they smile. Their clothes are mostly hides, of all sorts, but most frequently of sealskin in the summer and reindeer hide in the winter. Their rain clothing consists of bear and seal intestines, sewn together and made like one of our (Norwegian) homespun farmer's jackets, with a hood sewn on for covering the head. These raincoats looked as light and transparent as a piece of crepe. If one is to ask the Eskimos if these are good for rain when they have them on, one must not think that they say "no," but rather "Nak-gorok" (good) *[=naḳuuruq]*, although they seem no more to hold out the rain than would a crepe coat.

The steamship *Del Norte* fortunately arrived safely, and some of our companions were on board. We did the same imploring and begging for food from Dr. Jackson as yesterday. We received two pounds of butter, two pounds of sugar, and a little roasted coffee; we had some soggy crackers left over from yesterday—and this was to be nourishment for forty people for an entire day. Many wanted to have money from Dr. Jackson in order to buy themselves food, but he did not allow that, either.

## July 27

— During the day a group of us went to the marsh to find cloudberries, which we then went around eating. Dr. Jackson beckoned and waved to us that we should come to the hut, whereupon we went down. He wanted to tell us that we were to travel in the morning to Unalakleet, where we would stop over and have plenty of everything and be as if we were at home. He possibly thought that we would not ask him for more food, since we are going to leave in the morning. He said also that all of the supplies had been sent to Unalakleet, but we knew there were two large general stores and many large warehouses in St. Michael, full of as many sorts of foodstuffs as one could think of. But he let us go without, rather than to buy from the companies; so we headed for the cloudberry marshes. We also had a visit from our companions who had come with the *Del Norte*. The priest who was going to St. Lawrence Island came and saw to us—he visited Otto Leinan, as his wife was ill. Leinan had asked Dr. Jackson for a little money to buy some food for her and for the children, but Dr. Jackson managed to listen only rather coldly to Otto's request on behalf of the sick woman without giving him a cent.

## July 28

— Dr. Jackson's words did not come true, so we thought that it was all just to save money or to avoid buying us some food. But as well as we were thinking we knew his intentions, we yet had to go to him and ask for supplies. In such dealings, it was most often Magnus Kjeldsberg, and sometimes Thoralf Kjeldsberg, who had to be our spokesman; so it was this time, as well. They said to Dr. Jackson that we needed some supplies. Jackson said that we could have what we wanted and told them to write up on paper what they were requesting. Then we thought it would go gloriously, if we could get everything we would put down on the paper. So Magnus wrote on a small piece of paper: 20 pounds of rice, 4 pounds of butter, 4 pounds of sugar, bread and 24 pounds of meat (canned meat). The slip of paper was then taken to Dr. Jackson, and three men went along to fetch the supplies and carry them to the hut where we lived. When the boys came with the supplies, we found to our dismay that Dr. Jackson had permitted us just half of what we had asked for, which was even far less than what we needed. He let us have ten pounds of rice, two pounds of butter, two pounds of sugar, a crate of mildewed crackers, and twelve pounds of meat. We must divide this among forty-eight people, and anyone can figure out how much each one received for a whole day's nourishment. Since the cloudberries were now all picked in the marsh, both by us and by the Eskimos, there was nothing else to turn to. In the afternoon (after 12 o'clock), we saw Dr. Jackson and a woman walking towards the hut. When we saw that they were heading for our box, some of the boys took the crate of mildewed crackers, where there was nothing but mildew to be seen, and propped it right up against the door. When they arrived, Dr. Jackson with a frown looked in through the door. Next he saw how the woman noticed immediately the mildewed cracker crate, and she commented upon it to him, whereupon he also saw this. In order to put everything in a better light, so that she would not understand what our conditions really were, he said [to us] "You must not eat these, otherwise you can get sick." He must have known that we had nothing much else; but she saw it and commented to him, and he had to say something.

## July 29

— The entire day went as the previous ones, by trying to wait until we get to leave, for we realized that it would not be many days before we went to Unalakleet, where we were to receive compensation for everything for the duration of the trip so far, both to St. Michael and elsewhere on the journey—according to what Dr. Jackson had promised us.

## July 30

— We finally received orders to be present at the pier, and the boats from the *Del Norte* came and picked us up. We departed immediately after boarding. There are many people on board who were going north to look for gold.

## July 31

— Sunday. Early in the morning we anchored at Unalakleet. Since the weather was so good, Dr. Jackson ordered us to start unloading, even though it is Sunday. When we had come ashore, Wilhelm Kjellmann, Dr. Gambell and Regnor Dahl came from the mission out to the beach where we landed and bid us welcome. We heard that the schooner *Kenney [Louise J. Kenney]*, on which Regnor Dahl, Dr. Gambell, and the Lapps traveled from Port Townsend, had arrived safely at Unalakleet two days before. We worked all day at carrying supplies ashore, including those that belonged to the expedition and those that went to the mission, as was Dr. Jackson's wish. The landing place where all of the goods were brought ashore was about two miles from the mission; we were going to that area since a reindeer herd was there by the shore. The herd was cared for by two Lapp families, who had come to Alaska with Wilhelm Kjellmann five years earlier. They had been stationed at Port Clarence; but since Kjellmann had selected Unalakleet as a better place to have a reindeer station in 1897, it was moved to this place. During the winter, while Kjellmann was in Norway, these two Lapp families and a Norwegian boy—and also an Eskimo family and Dr. Kittilsen—had moved with the herd from Port Clarence here to Unalakleet.

The steam schooner *Del Norte*. *Photograph courtesy of the University of Washington Libraries, Special Collections, Negative 7723.*

## August 1

— The Lapps who had come from Port Clarence were Johan Sp. Tornensis, Mikkel Nakkila and Frederik Larsen, and their families; they said that gold had been found at Golovin, that there were many who had staked claims, and that each of them also had his own claim. Jafet Lindeberg heard this, and he was disinclined to go to Siberia and refused to go. At this, Dr. Jackson summoned Lindeberg, to settle up. In the meantime, we continued to bring goods ashore, and we finished after dinner. After that, the *Del Norte* began to steam northward. Jafet Lindeberg went with her, to go ashore at Golovnin Bay, as the ship was to call there in order to unload some goods for the mission. Frederik Larsen was born to Norwegian parents in Alten, and when Wilhelm Kjellmann was in Norway in 1894 to call for some Lapp families to go to Alaska on behalf of the American government, Larsen was able to sign up, and he went, although he was only sixteen years old. While Kjellmann was in Norway getting together the present reindeer expedition with which we went in the past winter, the winter of 1898, Frederik Larsen had married an Eskimo girl, which Kjellmann did not know about until he returned to Alaska with us, to Unalakleet. Since I have thus spoken of a marriage between an Eskimo girl and a Norwegian boy, which was apparently not done out of compulsion but rather was by free choice from both sides, I mention a bit about what I have been told about the Alaska women. Their worth among the Native people could be most clearly seen when one heard a description of their marriages.

Already from birth the woman is considered inferior to the man, and they have been bought as wives by anyone at all; so no matter how ugly, aged or infirm a fellow might be, he could get young girls as wives if he desired. The girl herself had no word in the matter, no will to assert. Apparently she was completely a slave in this business, for old, worn-out men who had old wives could divorce themselves from them and take new wives instead. So it would seem that when boys were ready to take wives, they had to be content with divorced women and widows. We heard that polygamy was frequent in Alaska, that a man could have ten wives at one time, but that he could divorce them and take new ones. So it seemed that he could have as many as he wanted. We heard that sometimes when a girl became ten or twelve years old, her mother would keep her in a darkened room for several months, so that the girl's face became white. She was not allowed to go out during the daytime, but only at night, and in her mother's company. Nor was she allowed to meet any men during that time. When the time was right, her parents sold her to the highest bidder, regardless of who that might be. Such crude and unenlightened customs are now almost completely abolished. Missionaries have worked against such practices in many places where missions have been established. Much has been done for the enlightenment of the native people of northern Alaska, where mission work has gone on for only ten to twelve years. But before the missionary began his work in Alaska, the Natives' religious practice was nothing but witchcraft and delusion. We thus heard how the Native parents often handled their daughters in a very wrong and inconsiderate manner, such that even their weddings would have been anything but joyful celebrations for Alaska women. One can wonder, if that is the way their happy family celebrations were, how must have been the sorrowful events?

The blame in this mistreatment of women in Alaska can be attributed as much to the white people, from earlier times, as to the native people themselves. These white people ("Civilized") conducted trade with the Natives and treated these defenseless women unscrupulously, buying them from their parents for indecent purposes. When such purposes were accomplished in one place, it is said that they moved to another location and renewed their shameful dealings.

## August 2

— We now have Wilhelm Kjellmann as our leader. Regnor Dahl no longer had anything to say, or any power. Dr. Jackson went northward with the *Del Norte* and will not see us further this year, since he will go down to the United States, after his return from Siberia.

Wilhelm Kjellmann organized the people into several groups, each group with fourteen men, not counting family members. Tents to live in were received; and we were placed into groups of four or five for each tent, according to how large the tents were. Kjellmann arranged for distribution of supplies in such a manner that it would be most convenient to apportion the allotment when provision day fell (once a month). He would apportion the distributions himself, with the help of a few men. The supplies were flour, meat, butter, coffee, brown sugar, milk, canned meat (roast beef), baking powder and yeast. All of these items we received, and enough this time, so that there was none who complained; rather, each accepted accordingly what was allotted, which was adequate. After the distribution of supplies, the question of cooks came up, as to who wished to be a cook. After a vote by the Norwegian group, I was made one cook. Johan Petter Johannesen became cook for group two, and Amund Hansen became cook for group three. The cooks were to prepare food and to take care that the portions lasted until the next provision day. Otherwise, there could be an uproar in the camp, for there would be no more supplies until the next distribution.

## August 3

— I looked about in the town, which consists of the mission house, the school, and twenty to thirty Eskimo huts. Also a Norwegian man by the name of Engelstadt lives there; he does business with the Eskimos. These altogether make up the entire town.

Unalakleet is a lovely and pleasant place in the summer, with a view over the sea and land to several miles' distance, while the mountains rise beyond, and forested river valleys are formed. In addition, the town is situated at the mouth of the river, called by the same name, which makes the location quite excellent; for the Eskimos can fish in the summertime for salmon, which is their primary occupation. Since the Eskimos live at the mouths of the rivers, one can understand how they are able to get their living—from the river. They also hunt and catch seals, but that, I learned, is done only in the spring and the fall. In the middle of the winter they do not catch anything; they consume what they have acquired during the summer. Their winter diet is primarily dried fish and seal oil. The men frequently bring in the fish and seals; in summertime the people have to pick berries wherever they are found—cloudberries, lingonberries and crowberries; these three kinds grow in northern Alaska—and store them for winter. In order to distill those to make fine dishes, they dig a hole in the ground, take salmon roe, and put the salmon roe and cloudberries together into the hole and cover it with branches and sod. They let it lie there until fall, when the ground becomes frozen. Then they dig up their "jam" and eat it frozen. Additionally, the Eskimos eat muskrats and mice—in short, everything, including the entrails of animals with digesting food within can be eaten by them; but they first let these things freeze. The same applies to raw fish: they permit it to freeze, then they dip it in seal oil and eat it; so from what we hear, it appeared that there is nothing that they do not eat. They have such a strong constitution that they can even eat the thick sea kelp, which not all creatures can do. The Eskimos eat that as if it were kohlrabi. At missions such as at Unalakleet, the missionaries have done much to enlighten the natives. The first missionary to begin his work here was Axel E. Karlson. He came twelve years ago, and has kept at his work faithfully through adversities and good times, and he still works as a missionary.

## August 4

— Wilhelm Kjellmann had selected the site where the reindeer station should be constructed, a distance out of Unalakleet, about eight miles from town. We were supposed to go up the river to get there, but we did no traveling today because there was a storm.

## August 5

— We had good weather and a good wind up the river. We placed the supplies, the tents and our clothes in the boats (the Eskimos' skin boats), and then we sailed upriver with three boats and fifteen men. Kjellmann was with us, too, as well as Regnor Dahl and Alex Joernes, who is to cook for Kjellmann. (Joernes is a relative of Kjellmann, and he joined the expedition while we were still at Haines Mission; he had planned to take the route over the mountains from Skagway to the Klondike, and Kjellmann took him into the group to go to northern Alaska together with us.) Kjellmann in his lighter boat arrived first to our destination upriver. So, before we others arrived, he chopped the branches off of a tall spruce tree and raised the American flag to the top. When everyone had arrived and had unloaded the goods from the boats, we all sat together in a group. Then Wilhjelm Kjellmann passed the bottle around, and thereupon everyone shouted "Hurrah" for our safe arrival and for our new home, Eaton Station, which we would now establish and build. Afterwards, Regnor Dahl and two others went downriver with the boats, to bring others up in the morning.

## August 6

— Sunday. We got to observe Sunday quietly—enjoyed ourselves by the bushy river banks, in clear, warm, sunny weather. We contemplated this spot where we had arrived: what good it could offer us; why we were brought from Norway; and what Kjellmann had found up here, among the desolate, uninhabited hills by the banks of the river. That he should go just here, about eight miles from the sea, into the woods, to construct the main station for raising reindeer in Alaska, is something for which I still could not find a reason. The only thing I can think of is that enough kindling and timber are present in order to have houses for the winter—which could perhaps be a good enough reason to construct the station here. And with regard to food for the reindeer, there was moss everywhere one looked; but there could be many other reasons that I do not understand.

## August 7

— Regnor Dahl had come up with group number two yesterday, so we were twenty-five men. Each has got his own job: some to clear away the forest, and others to burn off branches and all sorts of debris, in order to make clear and clean the area below the hill where the houses are to be built.

## August 8

— Regnor Dahl came upriver with the third group, so now everyone is present, except for some men whom Wilhelm Kjellmann had sent west to Port Clarence, to be on hand to receive reindeer that Dr. Sheldon Jackson had ordered and purchased in Siberia. The deer were to be brought over on the steamship *Del Norte* to Port Clarence, where the Lapps sent there were to take charge of the animals and watch over them, and were to bring them to Eaton Station in the winter. Jakob Haetta, and his wife and children, were also sent north, to Cape Prince of Wales; it was said that some reindeer that needed care were there. Now the entire crew at last has come upriver to Eaton Station. Everyone was assigned work by the manager Wilhelm Kjellmann, so that each has his own job and place to be: some to dig out lots and others to saw logs; some to carry the logs and others to transport goods from the river mouth, under the command of Regnor Dahl. Working time is from 7 o'clock in the morning until 6 o'clock in the evening, with one hour for dinner—a ten-hour work day. The women's work was to cut sedge and to dry it for use in the winter, when we shall put it in our Lapp-style moccasins to keep the feet warm. Some are to sew Lapp moccasins and Lapp-style boots, for winter use.

## August 9

— After dinner our manager permitted Magnus Kjeldsberg, Hilmar Hansen, Thoralf Kjeldsberg and Otto Greiner to go with nets to the river, to fish for salmon. They came back in the evening with a boatful of fish, 145 large salmon. We prepared large kettles of fresh salmon for supper, instead of cornmeal porridge.

## August 10

— Everyone worked, each at the post assigned. Our manager has an enormous task in hurrying from one work area to another, to see that all were at the job and that everything was underway. Johan Sp. Tornensis and Mikkel Nakkila, who had been guiding and watching over the reindeer herd, were given a vacation, and Johan Isaksen Tornensis and Aslak Aslaksen Gaup replaced them, along with the Eskimo Donnak and his wife, and an Eskimo boy, Martin. Together they are to take care of the herd, with Johan Isaksen Tornensis as their leader. Later on, Kjellmann sent two more Lapp boys to be of help with the herd, and a Lapp girl who is to be the cook for the group. The reindeer herd previously has been kept by the seashore, but today the herdsmen brought it up to the other side of the river (from our station) and set up camp there. According to Johan Sp. Tornensis' count, the herd numbered 150.

## August 11

— The manager let some more of the fellows go out to fish. Within a few hours, they had caught many large salmon, of which most were cut up and salted down, as salt fish. Sunshine and clear weather all day.

## August 12

— Everything fine and good. The sun was so warm that the crews had to force themselves to work. The lots were made ready for four residences, which are to be situated along the slope of a hill, with 100 yards between each building.

## August 13

— The first logs were laid down for the construction of four log houses, each one twenty feet long and sixteen feet wide.

## August 14

— Sunday. Since all had worked hard all week, everyone enjoyed Sunday, for we have the day off to rest from the week's labors.

## August 15 to August 17

— Nothing special happened, and everything was well and good. And indeed there was no time for anything but our work, at which we were all day. We had the evenings and nights to rest. Good weather, although at times there was some rain.

## August 18

— Wilhelm Kjellmann received word from Dr. Jackson that Kjellmann was to go to Unalakleet, in order to discuss some things before Dr. Jackson traveled to the United States. Dr. Jackson has come back already from Siberia, on the *Del Norte*, and he was on his way to St. Michael. When Kjellmann came back to our station, we got to hear the news from Siberia: the manager of the government's station there had run away, the warehouses were ruined, and several hundred reindeer that had been purchased by the U.S. government had dispersed and wandered away. Dr. Jackson was able to retrieve only a small part of everything there. The manager even had taken some of the stores, worth 20,000 dollars; he was presumed to have gone north with a whaling ship. There was no agreement as to reasons for all of this. When Dr. Jackson found the station in this condition, with everything in disorder, he took all of the remaining reindeer aboard the steamship; the people helped him with those that were caught. When the ship came to Port Clarence, the deer were put ashore. Ole Krogh and Johannes Rauna were no longer needed in Siberia, so they accompanied Dr. Jackson back to Unalakleet and came to Eaton Station.

From Wilhelm Kjellmann's former station came the news that his father, Torvald Kjellmann, had died on the 27th of May 1898.

We heard also a remarkable story, which is supposedly true, from the vicinity of Golovnin Bay, about three gold prospectors who lived together there. One lovely day, when one of them returned from the field, he told his companions that he had discovered gold, so they all headed for the spot. Since it was far to walk, they had to spend the night out, in the field. Whilst they slept, the one who had said that he had found gold, got up unnoticed and without the knowledge of his two companions, and sneaked down to the place where they had their supplies, presumably to steal them for himself. In the meantime the others had awakened and noticed that their third man was gone. They thought that something crooked was going on, so immediately they hurried down to their camp and found there the third man, who was preparing to leave with all the supplies. They captured him and sentenced him to death. But they said that they would spare him if he would show them where the gold was. So they set out once more. But again he ran away during the night, and he was captured in the same way. This time they took everything away from him and left him almost naked, whereupon they drove him off into the forest and told him that he could not come back or let them see him; otherwise, they said they would hang him—which they probably would have done, except that it is unlawful to execute anyone. The two were going to travel several miles from the spot where this took place, where they had threatened to hang him, so they left him out there to stay alive.

## August 19

— The boys once again received permission to take a fishing trip to the river. They came back with 150 large salmon, which they had pulled ashore with a small, ragged trawling net. Good weather.

## August 20

— The work progressed nicely onward. Sawing, chopping and carpentry were to be heard from all directions. Wilhelm Kjellmann went up the river with men to see if there was much timber close to the water so that logs might be taken easily to it and floated downstream to Eaton Station. He came back in the evening, having found sufficient timber. Good weather.

## August 21

— Sunday. We rested from our work. Sundays are observed here, although there is work enough for every day—just as at Haines Mission, where we did not observe Sundays. Here, that Sundays are observed is probably for the sake of the Eskimos, and also so that people could rest from the heavy labor. The weather is rather good, with rain.

## August 22 to August 27

— The work of building houses continued steadily and four log houses were completed. On all of them, roof beams were placed, but no roof. As soon as the beams were situated, Wilhelm Kjellmann presented a ridgepole can (a flask of whiskey) to each group that was building the houses. The weather was not so good this week, as it rained a lot, but it is not cold.

## August 28

— Sunday. The day passed by in peace and calmness, since we have nothing for amusement. So one has to sleep or, more correctly, rest quietly. The weather is also poor, with rain.

## August 29 to August 31

— Some of the men had to put roofs on the houses; as well, they installed inner walls of boards, and interior walls to divide each house into two rooms; also they installed the windows and put down the flooring, and placed in stoves. Others in the meantime were busy cutting wood and digging out the lot for the main building. Construction of that building is to begin just as soon as the four houses are finished and put in good order. When timber became short in supply, some of the men went upriver, cut timber in the woods, and floated it down to the sawmills at our station. Several men dug and formed a plot for the general store, which is to stand closest to the river, on the slope of a hill. On the long side, facing out to the river, the store-building is to be constructed on pilings, while the foundation of the other side, facing the upper slope, is to rest on the ground. A rather large quantity of goods has already been brought here, so that the store's inventory was nearly full, and in order that these supplies should not be out in the rain and be ruined, the store is to be built as soon as possible. Regnor Dahl with several men in two boats, brought supplies upriver every day. The weather has been humid.

We work steadily and hard, especially with the realization that in the fall, soon, about thirty men are scheduled to arrive; they would then go to the Yukon River and using reindeer would attempt to get the mail from St. Michael to Dawson, and from there to the United States. Their leader is to be an American, Richardson, who is in St. Michael. We are to accomplish as much as possible here, before people left. For the work concerning the mails, Regnor Dahl had been hired by Dr. Jackson while we were in Port Townsend, evidently. Dahl had said, on behalf of the Lapps, that they wanted him as their leader on the postal expedition in Alaska, and that everyone was pleased with his keeping of law and order in Port Townsend, as no one said anything against him there. We only later heard of this and that Dahl was hired as leader under Richardson.

## September 1

— Wilhelm Kjellmann sent some men to St. Michael with the mail, and to bring mail back.

## September 2

— News came from Jafet Lindeberg. He was at Golovnin Bay and had already staked three gold claims in the Eldorado District, near Council City, where they had found good prospecting (rich soil). We all began to think how one could get there, for possibly others could get claims. But that was not so easy, since we are in the employ of others. The whole matter rested on what our manager, Wilhelm Kjellmann, would be able to do about this. We also have heard that Mr. Richardson, who had agreed to deliver the mails over to the Yukon Territory, has grown fearful and presumably has broken his contract. Therefore, those of us who were to go with him would probably not go; our minds began to call us northward to the gold fields.

We also have heard news about the schooner *[Louise J.] Kenney*. It is reported to have sunk off Cape Prince of Wales, but everyone on board was rescued.

The weather is stormy, with rain.

## September 3

— Everything fine and good. We wait a bit for the mail-crew and think about the gold news that had come. More or less good weather. Rations were passed out, this time less.

## September 4

— Sunday. The day passed by quietly, as usual on Sunday here. Rain and fog all day.

## September 5

— Everyone is at work. A log house the same size as the others, 20 x 16, was begun in the woods. After it has been finished, it is to be dismantled and carried up the hill and raised in a row with the other houses. The weather is good, with sunshine.

## September 6

— Since there is good weather, we thought that the mail would certainly arrive today, but it did not come.

## September 7

— The boys finally arrived from St. Michael, bringing many letters that had come from Norway. Good weather.

## September 8

— Several hundred meters up the river, a sawmill (manual saw) was started, and some of the men cut timber to carry to the mill. Farther upriver, another group of men is cutting trees and floating them down to the sawmill at our station. Rainy weather.

## September 9

— Dr. F. Gambell received one of the completed houses to live in. The other houses still stand empty. Everyone must be content with our tents, until there are enough rooms so that all can move into the houses, in order to avoid dissatisfaction. Many wanted to move indoors, out of the tents, but this is not permitted. Poor weather with rain.

## September 10

— The doctor has made arrangements for some of the boys to join him one hour each evening, after finishing work, in order to learn English. School is to begin on Monday and to continue each evening except Saturday and Sunday. Good weather.

## September 11

— Wilhelm Kjellmann gave permission to some of the fellows to go to Golovnin Bay in order to stake gold claims, both for themselves and for those of us who will remain at the Station. For this purpose, they received a written authorization (or power of attorney).

## September 12

— The men could not go, because of a storm, so they have to wait until tomorrow.

RESIDENCE OF THE PHYSICIAN AT EATON REINDEER STATION, ALASKA.
Photograph by F. H. Gambell. M. D. (See page 12.)

At Eaton Station, 1898.

*Photograph published in Sheldon Jackson's report, 1898b.*

## September 13

— The storm still hindered those who are to go north to Golovnin Bay.

## September 14

— There is good weather, so the lads (Regnor Dahl, Magnus Kjeldsberg, Wilhelm Basi and Ole Krogh) could leave. They are to be away three weeks, and then return by boat. Along with them went Johan Sp. Tornensis and Mikkel Nakkila, who have received permission to remain at Golovnin Bay during the fall in order to do assessment work on their claims, so as to be able to retain them during the next year. (A legal requirement was that one must work a claim in order lawfully to keep it, and so that no one will be able to jump his claim. A further requirement is that a person who takes one or more claims must do one hundred dollars' worth of work each year, on each claim, if he wants to keep the claim.) Johan Sp. Tornensis and Mikkel Nakkila are to return in the winter, after the first snowfall.

## September 15

— A group has an opportunity to go to St. Michael, to take our mail from Eaton Station, and try to make it in time for the mail to be sent to the United States by ship. These presumably would be the last letters we should be able to send by ship in 1898. Since fall is beginning to approach, with frost, the ships have to leave before the ice comes, in order not to be frozen in. Clear weather, but a little cold.

## September 16

— The roof was put on the general store, and we began to carry in the supplies.

Some of the men worked at breaking up stones, up on the hill, for a foundation wall for the main building. Otto Leinan, who is a mason, was placed in charge of the masonry work. The weather is fairly good.

## September 17

— Everything is good and well, in a sense, except that many are complaining that they are ill due to the hard labor. No particular attention was paid to the complaints, for the work absolutely has to be done in order to have housing for everyone in time for winter. Nothing has come of the trip to the Yukon Territory that had been planned. As far as I can make out, that may be related to a matter of the supplies, in that we received less at the last apportionment than what we had received when we came here. And now, since there are more people here than the number for which the station had been planned, it seems that supplies are not sufficient for that trip. Rolf Wiig has been placed in charge of the general store; the people are to obtain what they need from him. Everything that is taken is recorded. Not yet is it clear whether payment for things received at the store will have to be made sometime in the future, though I think it likely that we should have to pay for some of them.

## September 18

—Sunday. The day of rest was a precious day, as it is always shorter than the others. Kjellmann sent two of the lads up the river with a boat to help two gold searchers with their supplies. Their camp was about seventy miles up the river from Eaton Station, and they were going to try to find gold in that area. The weather was good, but cold when the sun went down.

## September 19

— As usual, Wilhelm Kjellmann was up early in the morning and turned everyone out for work. If anyone was not out of bed, he had to hurry his feet out of the tent. Many are complaining that they are ill, and when Kjellmann asks where it hurt, the Lapps reply "I have a sore back," but that did not help matters much. Everyone must turn out for work, without exception. But when one could see that a man was indeed sick, then that man received rest and any needed care by the doctor.

## September 20

— The foundation for the main building was completed. The weather remains good.

## September 21

— Wilhelm Kjellmann gave everyone his assignment, and those who were relatively the best builders began construction of the main building. Kjellmann himself had drawn on paper the building's dimensions and style. The size of the house was forty by thirty feet, and it is to be two stories in height, with six rooms downstairs and five upstairs. The weather is unchanged.

## September 22

— Some log cabins that had been under construction were completed; the logs have been marked and the cabins dismantled; each has been brought to the site of its lot and then erected in line with the other houses, up on the ridge of the hill. The weather was cold and foggy.

## September 23

— It snowed today for the first time, but the snow was melted away during the day by the sun. However, it was very cold in the evening after sunset.

## September 24

— The boys returned from upriver where they had helped the prospectors in taking their supplies to their camp by the river. As far as the boys had learned, the prospectors had not yet found any sign of gold. Wilhelm Kjellmann reduced the time of our dinner break to half an hour and increased the supper time by as much—for it is certainly obvious to him that a great amount of the work here has not yet been completed and that winter is rapidly approaching. There is now half an hour for dinner, and the men are to quit work in the evening half an hour earlier than before. Also, as it is beginning to get dark before quitting time when an hour is given for dinner, the new schedule is better. The weather was a little milder today.

## September 25

— Sunday. As usual, Sunday passed in all peace and quietness. The weather was calm, with fog mixed with rain.

## September 26

— I could not find anything unusual or new going on. Everyone went to work as usual, after Kjellmann had called them out. He was constantly keeping track of the minutes, every single morning. The weather was not particularly good.

## September 27

— The boys who had finished cutting timber and constructing the log house were sent by Kjellmann back to the woods, to cut trees and build one more of the same size as the others. It is to be ready for people to move into, for the winter. The weather was good.

## September 28

— Kjellmann met with the Lapps who were complaining that the work was too hard (which was in fact the case); they said that they had sore backs. Kjellmann sent them out into the hills to gather moss for winter feed for the reindeer. The weather was humid.

## September 29

— Poor weather and very cold, but everything good otherwise.

## September 30

— There is nothing going on but the same hard labor—nothing for amusement. Everyone has been too tired to go to evening school to learn English, so the doctor finally had no one to instruct. This evening, however, he had two who were continuing, but perhaps they too will become tired.

## October 1

— When I thought about home and compared our Saturday here (today) with that of the folks at home, I found differences not in just one way, but in all ways, as anyone would discern in comparing the monotonous, lonely and empty Alaska where we are, with the activity that takes place in churches at home on Michaelmas (St. Michael's Eve) (Saturday). There are great differences. It was lonesome and dreary on this Saturday and little went well or quickly as we struggled with our labor from morning until late evening. But good weather.

## October 2

— Sunday—St. Michael's Sunday.The morning dawned with clear weather and sunshine, but the sun is not giving much warmth. Sunlight has become weak and faint for Alaska, so we begin to expect winter, which will soon arrive.

After dinner, in the afternoon, the boys returned from Golovnin Bay and brought us news from Council City; already many have lost out for a claim. Jafet Lindeberg was among those who had good claims, but the boys from Eaton Station came there too late. All of the places where gold had been found had been taken, so they got only a few claims in the outer area, which no one earlier had wanted to take. They recorded altogether eleven claims then; some of them tried to prospect, and they found a few bits of color (gold), so they did not know whether there was much of any gold or not, in these claims. The best news for us is that much work would be done at the gold fields next summer. So we all planned to take the option of working there, if we did not have claims ourselves, and probably thus earn many times more than what we should earn in the service of the government. A whole group of us younger men planned to resign next spring and travel over there to work. Olai Paulsen came to our station with the same party. He had received tickets from Dr. Jackson to travel from Seattle to St. Michael, after he had been home. Paulsen had left Seattle long after we had departed; when he arrived at St. Michael and could not find transportation from there to Unalakleet, he went on a small schooner to Golovnin Bay (where he knew a Swedish mission was located), in order to get from there to Unalakleet by one means or another, depending upon what transportation he might find. While he was waiting for transportation, he had the opportunity to travel with others to the gold-district, where he staked a gold claim.

## October 3

— Thoughts of gold, and of our supplies—whether they would last through the winter—and of when to leave Eaton Station, last throughout most of the day. The work proceeded at top speed, with chopping and sawing timber and building with all our might. Primarily, the work was on the main building, for it most of all had to be completed and ready before winter's cold set in with frost that would hinder the work. The weather was still more or less good, although already it is quite cold.

## October 4

— Kjellmann had some of the men begin to dig a road along the slope of the hill, below the houses, before the ground freezes. The weather was good.

## October 5

— The weather is unchanged. Many of the Norwegian men resigned from government service, after having spoken with Kjellmann with regard to the supply situation and the journey to the gold fields and at what time of the winter they should depart. Kjellmann forbade no one's wishes; everyone was permitted to do what he thought best.

## October 6

— Nothing unusual, everything well and good. But Klemet Nilsen, from Karasjok, is sick. He is lying in bed in a tent. Storm.

## October 7

— The weather is good. Everyone has been busy with his usual work, according to his assignment.

## October 8

— Good weather, but otherwise cold. Work has gone well the past week. The main building already stands quite high, and presumably it will be completed to full height by next week if the weather remains good.

## October 9

— Sunday is always a welcome day (as previously mentioned), as one can rest from work; and the weather was good today.

## October 10

— Kjellmann sent several men to Golovnin Bay, to stay until there is snow, then to come back, around by the land-route, with some reindeer that the mission had obtained on loan from the government. (Kjellmann wanted the reindeer to be at Eaton Station.) A couple of men went along to return with the boat.

## October 11 to October 14

— Work progressed steadily, getting sufficient timber brought in and sawed so that the builders should not have to wait for materials. The men hurried with all their strength, and the sounds of pounding and carpentry could be heard day in and day out. Moreover, the weather was still relatively good.

## October 15

— Our main building has progressed to such a height that the top logs of the long sides were to be put in place. A smaller log cabin was finished also. The weather is very cold.

## October 16

— When I went to fetch water from the river this morning, I was astonished to see that the entire river was full of ice, which floated downstream. As far beneath the surface of the water as I could see, ice lay on the stones, down to the river bottom. The ground and all of the ponds are frozen. Although there was clear weather and sunshine all day, there was no melting; now the sun no longer gave us warmth.

## October 17

— Earlier, in the summer, one did not have to work long before one became too hot to labor; but now one has to be warmly dressed, with mittens on the hands, and one has to work hard in order to keep warm. Although we still do not have winter with snow, at least we have winter with cold.

## October 18

— During the night, the river froze over, so that one can walk on it already. It was so cold during the day that the Lapps had to get out their winter clothes. They complained a bit that it is terribly cold and they went now and then to Kjellmann and asked to move inside, but they were told only "no."

## October 19

— The weather was a tiny bit milder when the day began, and it moderated a little, so that the cold was not so penetrating as yesterday.

## October 20

— We had a conflagration. It happened that Peder Berg, who has sprained a foot, was not able to work and had to stay in his tent. Since I needed someone to help saw wood, Peder wanted to lend me a hand. He did not want to extinguish the fire in the stove in his tent, and while he was out, the canvas caught fire and in an instant everything was in flames and became reduced to ashes. Those who lived in that tent lost all of their clothing—all burned except their suitcases, which were rescued from the flames. Fortunately, the tent stood sufficiently far from the other tents so that none of the others was damaged.

## October 21

— The main building received its ridgepole, and part of the floor was laid.

Olai Paulsen became the smith, to make stovepipes for all stoves—which are to be in all of the rooms of the building.

## October 22

— Work continued at roofing the main building; but the weather was biting cold. A steady, constant wind blew along the river and made the air very cold there by the river banks.

## October 23

— Sunday. Very cold. There is skating on the ice for the Sunday amusement, as today there is nothing else to occupy our time.

## October 24

— Many were ill, but all those still on their feet and managing to hold out are at work. Mostly, the Lapps could not go to work; some of them went about from tent to tent, or kept in bed in their own tents, and froze. (Sometimes they went to the doctor and asked for medicine; when the doctor gave them some medicine which had a taste they did not like, they requested other kinds that would taste sweeter to them, or for "Nafta;" they liked that.)

## October 25

— Since the weather was cold, Wilhelm Kjellmann permitted some of the families to move into the houses, for the sake of the children. But all single men are to live in the tents until the main building is completed, at which time a number of other people are to be moved there, too. So we are to remain in tents and wait until there is room enough in the building for all to move inside at the same time.

## October 26

— Our rations were distributed for the next twenty-eight days. Rations were reduced again, for the second time. That is unpopular and led to an uproar in the kitchen tent at mealtime, for we see that the rations now received are far less than what Dr. Jackson promised in Port Townsend; at that time, the monthly allotments (for each group) were announced to us in Norwegian language by Regnor Dahl, reading from the list that Dr. Jackson had provided, and we had written down the items—40 pounds flour (wheat), 20 pounds crackers, 3 pounds rice, 3 pounds oatmeal, 8 boxes matches, one-half pound of soap, 20 pounds potatoes, 4 pounds butter, 1 liter syrup, 15 pounds meat [beef], 10 pounds pork, fish free, 1 pound coffee, 3 pounds tea, 3 pounds sugar. These were to be the rations (monthly) in Alaska after we left Port Townsend, and they were drawn up in a list so that we did not have to fear that we should not have sufficient food. Now we did not get much more than half of that, and in some cases, less than a third.

## October 27

— Storm and driving snow all day. Two prospectors, who were living about twenty-five miles upriver from our station, came in today and spoke with Wilhelm Kjellmann. One of them is a doctor by the name of Southward, and the other is an engineer, A. Spring. Both are from New York. They had come to Alaska during the summer and they are going to spend the winter in a cabin that they had built by the Unalakleet River.

## October 28

— The cold was so severe that it hindered much of the work. Those who had to be outside all day at hard labor then had to go at night into tents where they had no heat, nor were they able to dry their clothing properly. It is no wonder that many are sick and have severe colds. We now had no hope for those who had gone to Golovnin Bay, in bringing back the boat to our station, since the sea has frozen. They are to delay their return until snow falls and all rivers are frozen in order to travel around Norton Bay with the reindeer, in which case they probably would not return to our station before Christmastime.

The first overland mail passed Eaton Station in the afternoon, carried by two men, one of them an Eskimo, and eight dogs pulling a sledge, which also held their supplies and clothing. They are to deliver the mail to Nulato on the Yukon River, and then farther up the Yukon to Tanana. Since nothing had come of the proposal for mail delivery with Richardson, the contract to deliver mail from St. Michael to Tanana City each month was taken by Engelstadt in Unalakleet. He is hiring people to do each mail run.

## October 29

— Sunday. We spent the day by running from one tent to another, thereby passing the time and keeping warm, as it is very cold.

## October 30

— Some men were sent into the forest to cut trees for winter fuel. Cold, clear weather. The last day of October also is the same.

## November 1

— Clear weather, calm and frosty. In the summer, when the steamship *Del Norte* left Unalakleet, the schooner *Kenney* sailed north at the same time; several men from the Reindeer Expedition were aboard, to be put ashore at Port Clarence in order to gather up the reindeer that the *Del Norte* was to bring back. But when the *Del Norte* returned, those who had been put ashore at Port Clarence also returned, except for two who remained behind. Those two set out a few weeks later on foot, sometimes traveling with Eskimos who were going east, until at last they arrived at Golovnin Bay. There they were fortunate enough to catch a schooner going south to St. Michael. From St. Michael, they went to Unalakleet on foot, and then up the Unalakleet River to Eaton Station. During their travels, the two men were able to hear news from the gold fields, including that Jafet Lindeberg had made a rich gold strike near Cape Nome, about seventy-five miles west of Golovnin Bay. But they cannot give a complete and accurate account of the discovery of gold at Cape Nome, since neither of them is fluent in English language. Nor did they know how many had been together on the discovery journey. But we heard that in fact gold had been found; so we once again begin to speculate about our own chances.

## November 2

— All who are still healthy stayed at their jobs and kept at the work with all their strength, so that we can get out of the tents soon. Those who are ill are putting on all of their clothes to keep warm, as it is very cold now.

## November 3

— Wilhelm Kjellmann and Regnor Dahl went to the area where the reindeer herd is located, about four miles from the station, to find a place where the deer could be kept during the winter. They are to be away for several days. Weather unchanged.

## November 4

— Seven men—one Norwegian and six Lapps—are ill and unable to work. Some are said to have scurvy, but most have severe colds. Weather unchanged.

## November 5

— Kjellmann and Dahl returned from their trip into the mountains. Dr. Gambell has been called down to Unalakleet, as some people there are ill. Weather unchanged.

## November 6

— Sunday. Dr. Gambell returned from Unalakleet. Everything at Eaton Station is peaceful and quiet, as usual.

## November 7 to November 10

— The time was spent getting the main building finished.

## November 11

— Almost everyone was permitted to move into the houses; only group number one remains out in tents. We are obliged to wait until tomorrow, for the group will go to the second floor of the main building, and it is not ready. Wilhelm Kjellmann will live downstairs, while one room is to be the doctor's office and the schoolroom.

## November 12

— The last of the cooks moved in from the severe conditions of cold, to the second floor of the main building. One room there is to be used for all of the food preparation. The large hall is the dining room and also one group will live in it. Kjellmann has his room downstairs, and also the kitchen where Alex Joernes works. Dr. Gambell has his room on the second floor, next to the large hall, and he is going to prepare his own food. Clear weather, and much frost, today.

## November 14

— Now the streams, the river and everything were frozen, and so the Eskimos can drive their dog sledges wherever they wished to go. A whole group of them are at Eaton Station; they brought along some fish to sell. The fish resemble a sort of cod, about the size of an average herring, and they were caught at river mouths and along the shore of the sea, after the ice became strong enough for the Eskimos to venture out upon it. They fished for this cod by first chopping a hole in the ice; then they used a small stick of wood, to which was fastened a two- or three-foot-long line at the end of which was a small iron hook, like a fishhook. They lowered the line down through the hole in the ice, into the water, but they did not keep it still; rather, they jiggled it up and down until the little fish was attracted to the hook. Then the fish is cast out onto the ice, and the Eskimo would at once kill it by beating it with a small wooden club that was held in his other hand.

The women as well as the men go out onto the ice to catch these small fish. No matter how much frost and cold there is, they can sit out on the open ice all day. As long as they do not freeze, they do not have to move all day. When they get hungry, they take some of the frozen small cod and eat of them as they

Headquarters. Warehouse.

EATON REINDEER STATION, ALASKA.

Photo by Assistant Engineer A. C. Norman, R. C. S. Page 12.

Eaton Station, 1900.

*Photograph published in Sheldon Jackson's report, 1901.*

are. A mother can sit there and fish all day, in chill and wind, with her baby on her back. If the child begins to cry, the mother moves herself from one side to the other until the child becomes quiet. Thus they are fishing all winter long. Since great numbers of these small fish are present swimming under the ice from early fall until spring, the Eskimos have fresh fish all winter, even selling a large number of them, as they are doing at Eaton Station. They usually feed these fish to their dogs, which also require a lot of food during the winter. For winter transportation, an adult Eskimo might have ten or twenty dogs that he has to feed. I believe that if they did not have the little codfish in the wintertime, the Eskimos might have ceased to exist; for they are not always inclined to collect sufficient winter provisions during the summer when all the streams and rivers are full of fish. The men seemed to sit around by the river banks and sun themselves, while fish were in the water all about. The women, on the other hand, are more industrious—so it seems that, in a sense, Alaska women have to feed their husbands. In fall, after the ice and snow come, the women chop holes in the river ice in such places as are suitable to set nets under the ice, to catch salmon. After the ice becomes so thick that they cannot catch salmon, they chop through the ice several feet and place the nets again under the ice; but they cannot catch very much in that way. Ice several feet deep can form on the rivers in a very short time. The primary things that the Eskimos purchase are, first, flour, tea, sugar, tobacco, pork and ammunition; besides those items, they buy fabrics, especially calico, canvas and blankets. Most often they do business by trading goods for other goods. They bring in furs of all kinds, already sewn—fur coats and jackets, hide boots, and leggings, pants and gloves of fur—trimmed and designed after their own style. Their clothing is different from other fashions, even those of the Indians', but they resemble most closely the style and fashion of the Lapps, with pants and coats (jackets), although I consider that the Eskimos' clothes are more attractive than those of the Lapps.

The law in Alaska gives no one permission to sell alcoholic spirits to the Natives, nor syrup and a type of brown sugar (the people purchase those items only in order to make spirits, an art they presumably learned from the white man). It is said that the distillation of spirits was unknown in Alaska from ancient times until the year 1875, when a man from Sitka taught the native people the art of distilling spirits for themselves. After that came terrible, bad times for the Alaska Natives. Terrible misery in all respects occurred among the Natives after they learned to make spirits, which gave them time for nothing else.

## November 15

— Clear weather, much frost, so the men got a bit of a break in work. J. Hagelin, Mrs. Karlson and Miss Johnson drove up to Eaton, to see the new station.

## November 16

— Two men arrived from the reindeer herd, each with his own string of sledges full of meat (eight reindeer had been slaughtered, as some were of low vitality and some had broken legs). Work was partially suspended, so the men get to rest a bit.

## November 17

— Some worked at arranging and furnishing the main building. Some sawed planks and others chopped timber for firewood.

### November 18

— Everything is well and good. Those who were ill are recovering, but two of the Lapps remain quite ill due to their strenuous labor; they suffer from colds and weakness.

### November 19

— Everything was cleaned and washed for Sunday, in accordance with the rules posted by the manager. Now the job of cooking will be done in turns, each cook with a regular week, during which time only he will be preparing the food. The next week, another will take the job.

### November 20

— Sunday. We remained quiet and rested from the week's toil. The Lapps strolled a bit from one house to another, as is their custom on Sundays. The weather is clear and cold.

### November 21

— All went to work, each to his own task. Kjellmann gave the orders, mainly to saw wood, chop logs, and stack together in piles, which become huge. At each pile, sawing was to commence. Some men dragged logs on sledges across the ice and down to the station, where they are to be sawed into boards. Some of the men were sent to gather winter kindling and carry it to the houses.

### November 22

— We had news from the Yukon Territory, as two men came over the mountains from Nulato, on their way to Golovnin Bay to look for gold. They said that there were very bad times up in the Yukon, with little to earn and too many people. Many of the people were in a sorry state, sick and miserable, without any recourse. Many had died of scurvy and various other diseases.

### November 23

— It was made known all over the station that everyone is to observe a holiday tomorrow, as that is the American Prayer Day. No one objected; rather, we should gladly observe Prayer Day, as one was accustomed to such days from the Old Country. The weather was still unchanged.

## November 24

— Prayer Day. The day progressed as the other days, but in stillness and quiet, until towards evening in the moonlight, we saw reindeer coming, with bells jingling. Outside of the main building, they stopped. It was the lads who had been at Golovnin Bay; they had gone there on October 10th to fetch the reindeer, but they had not received any, for the mission there would not turn over any deer for the reason that Dr. Jackson had not paid for care of the animals over the past summer. The boys had been given only as many animals as they needed to take them back to Eaton Station. With them were Jafet Lindeberg and Dr. Kittilsen, who came on a visit, after having found a rich gold district where they had valuable claims. After supper, we began to hear news from the gold fields, told by the fellows who had been sent to Golovnin Bay. They told us how it happened that this rich gold district had been found at Cape Nome; Jafet Lindeberg happened to be there, and he discovered it. He found it and took the first claim. Thus it occurred, according to the fellows, and as Jafet Lindeberg told us himself. Olai Paulsen had heard the same account, from Lindeberg and Hagelin, regarding their first trip to Cape Nome last summer. The full account was that late in the summer (1898), the missionary Huldtberg, J. Brynteson, Hagelin and two others from the United States traveled north by boat to Cape Nome and sailed up along the river that is now called the Snake. They went as far as Mountain Creek and Rock Creek, prospecting, but they found nothing, whereupon they came down, crossing the mountains to Anvil Creek. Here they crossed the river, approximately where claim No. 3 Below is located. We heard that Huldtberg told it: "I was the last man to cross the river, and I tried to pan. I found some gold color. My companions had gone quite a distance ahead of me, and I didn't say anything about it right away." Hagelin waited longest for Huldtberg, while the others hurried on to get to the boat. They then sailed back to Golovnin Bay. In the meantime, Jafet Lindeberg had gone up the river to Council City, hoping to find a gold claim there. Olai Paulsen also wanted to go up and look around, and he waited for transportation, in order to go up the river. At Golovnin Bay, Brynteson and E. Lindblom met for the first time, after Brynteson returned from the Cape Nome journey.

These two then went into partnership and traveled to Council City in order to prospect. A few days later, Huldtberg, the missionary Anderson and Olai Paulsen, as well as a few others, went to that district, and there they all met. While they were there, they spoke about making a trip down to the mission (Unalakleet).

Huldtberg offered his place to Paulsen, saying that he did not have to go to Unalakleet, "for I know where there is more gold than there is in Council City." When transportation going east was available, Paulsen was more or less obliged to go, but he wished that Jafet Lindeberg could take his place here, and he therefore wrote to Lindeberg, giving him legal authorization to stake a claim for him. Then Paulsen went to Unalakleet and returned to Eaton Station. When J. Brynteson and E. Lindblom met Jafet Lindeberg up in Council City, and since Lindeberg had a large quantity of supplies that he had purchased in Seattle, they went into partnership—Jafet Lindeberg, J. Brynteson and E. Lindblom. After they returned to the mission (Golovin), they began to equip themselves. For their journey northward they obtained a flat-bottomed boat, to which they added a keel; they got some Eskimos to make a sail, loaded on supplies, and sailed away, the three of them together. After a few days they came to Cape Nome, where they stopped on the Snake River. From there, they went up to Mountain Creek, Rock Creek, Snow Creek and Glacier Creek. In all of those rivers they found relatively good prospecting. Then they went over the mountains to Anvil Creek, where they found good prospecting along the entire stream; but they found the richest signs of gold right at the opening of the valley through which the creek flows, where the mountains rise on both sides, about three and three-quarters miles from the mouth of the Snake River. There they staked the first claim, "Discovery," 1500 feet long and 660 feet wide. They gave the river the name of Anvil Creek. Next they staked out claim No. 1 Below, and Jafet Lindeberg took those two claims. J. Brynteson then took claim No. 1 Above, while E. Lindblom staked his claim about one and a half miles farther up, since he had found good prospecting there. Then they went across the mountain, and each staked his claim on Snow Creek, Mountain Creek, Glacier Creek and Rock Creek. They gave these rivers those names during their explorations, after which they crossed the mountain again on their way down. They also named the river that runs into the sea, calling it the Snake River. They then sailed back to Golovnin Bay, where they made arrangements confidentially with some of their best friends and borrowed a small schooner from the mission, sailing northward at last in the month of October. Seven men went, including Jafet Lindeberg, J. Brynteson, E. Lindblom, Dr. A. Kittilsen, Johan Tornensis, G. Price, and Konstantin (a young Eskimo from the mission at Golovin). They arrived safely at Cape Nome, where they stopped in the Snake River, left the schooner by the river, and walked up to Anvil Creek where they set up their camp. Kittilsen was named recorder, and he made up the laws for the district, which were primarily to regulate matters concerning gold-claims: how large a claim could be staked out; how recording must be done in order to retain each claim; that one had to pay two and a half dollars for each claim; one could stake a claim in each stream; one could not stake more than seven claims for another person without written authorization. The district was to extend 25 miles along the sea and 25 miles inland, and there was to be a Recorder of Claims for this district; all claims were to be recorded in the same district in which the claims were found; claims were to be recorded within forty days, otherwise someone else would have the right to take them. The size of a claim was to be 1320 feet long and 660 feet wide (20 acres). They stayed there for a month, working a wooden rocker (a special kind of apparatus fashioned from wood, with which one can wash out gold by pouring sand into it—gold may be in the sand; one man moves the rocker forth and back like a cradle, while the heavy gold sinks to the bottom and the sand goes out over the sides, with the water). They went over to Snow Creek also with such an apparatus and rocked out 1775 dollars altogether from claim No. 1 in that stream. Just to the east of Anvil Creek, they also prospected along a stream that runs out to the mouth of Snake River. Here they found good prospecting, staked several claims there, and named the stream Dry Creek. So they had located about fifty claims on the above-mentioned watercourses, as well as organizing the district and naming it the "Cape Nome Mining District."

Now they headed back; but since it was so late in the fall, the schooner became frozen in the ice of the Snake River. They had to make their return journey on foot, about eighty miles along the coast, and then to Golovin, where they made known all the finds. (News of the gold found at "Cape Nome" traveled quick as lightning, and soon people began to stream northwards from all directions—from St. Michael and from various places of the Yukon area. Jafet Lindeberg had staked claims for many of his friends, for example Wilhelm Kjellmann, Magnus Kjeldsberg, Olai Paulsen and Wilhelm Basi, while the others had staked claims for their acquaintances, for example for the missionary Karlson in Unalakleet, for the military doctor in St. Michael, as well as for a number of other men there, and for the missionaries Huldtberg and Anderson, as well as for Hagelin, who were the first to get claims in the Cape Nome Mining District by powers of attorney. Later on, many joined company and sent one or two men to stake claims for them with power of attorney; nearly half of the recorded claims evidently have been staked by the power of attorney as permitted by the district law, which was a paragraph of the United States mining law.)

Prayer Day that began with calm and quiet rest ended with our spirits lively.

## November 25

— Everyone was excited by the excellent news of the gold-finds, and all went about thinking solely about how to get to the gold fields so as to make a claim before everything was staked by others.

## November 26

— Wilhelm Kjellmann ordered a number of men to go to the reindeer herd to fetch thirty animals, in order to travel to St. Michael for the purpose of transporting some supplies back to Eaton Station.

## November 27

— It is quiet, and the usual work lighter, and all the while we listened to the reports about, and talked about, Cape Nome.

## November 28 to November 29

— Kjellmann and thirteen men departed for St. Michael with the reindeer to obtain supplies. Several of the group who went were going to apply for citizenship papers, for without such papers they felt certain they would not be able to locate on land or make gold claims. Jafet Lindeberg, Dr. Kittilsen and J. Brynteson, who had been staying at the mission at Unalakleet, were also to go to St. Michael; they wanted to put their affairs in order at St. Michael and also to carry news of Cape Nome. Kjellmann left word with us that as soon as he returned from St. Michael, he would prepare to go north with some of the men in order to stake gold claims.

## November 30

— Mail arrived from the Yukon, from places close to Nulato, destined for St. Michael. The weather had been clear all of the previous week until now, but the cold had increased greatly during this time.

## December 1

— It is quiet at the station and the work has nearly stopped. It was not good to stand outside in the formidable cold, and in the wind that blows constantly down the river and bites so bitterly one's nose.

## December 2

— Nothing special; everything is fine as usual.

## December 3

— Wilhelm Kjellmann and Jafet Lindeberg returned from St. Michael in the evening, with some strings of reindeer and sledges. Dr. Kittilsen is going to stay in St. Michael for a while.

On their trip to St. Michael, two men had been left behind on the ice, since their reindeer evidently were not strong enough to keep up with the main party. Those two were out there on the sea-ice, and the ice broke loose from shore, and the wind blew them out towards the open sea. Fortunately the ice stopped, but a long distance of open sea lay between ice and shore, and they were not able to get to the shore. They could only remain quietly on the ice and wait to be rescued. They stayed overnight in the cold and wind with no wood to make a fire, and the reindeer did not have enough to eat as they had no moss. For an entire day they were out on the ice; it became so cold during the night that more ice formed on the water, to a thickness they decided to try out, and they reached the shore. They went on to St. Michael and rejoined their companions just a little before the men were preparing to leave, to return to Eaton Station. The men formed two groups, one of the groups to follow later to the station. Regnor Dahl, who had also been in St. Michael, has been hired by the North American Transportation and Trading Company to go north to Cape Nome in order to stake claims by power of attorney on behalf of the company. One man is to accompany Dahl. Fredrik Larsen was hired by the company to watch over the reindeer.

## December 4

— Sunday. The second group did not return until dinnertime (about midday). Fredrik Larsen is among them, having left Regnor Dahl at Unalakleet. Larsen was not happy with Dahl and did not want to go with him to Cape Nome, for Dahl evidently had wanted to kill Larsen already during their trip from St. Michael to Unalakleet. Everything was not going the way that Dahl wished, whereupon he hastily went for his revolver. But fortunately he had left his revolver at Eaton Station, so he reached for an empty pocket.

## December 5

— Kjellmann announced that about twenty men were to prepare for the journey to Cape Nome on the 7th of December, and decided about which ones amongst the crew are to go. Lindeberg is to go along as he is familiar with the area.

## December 6

— Kjellmann sent men to the herd, to fetch about forty reindeer. The rest of us helped to get ready the sledges and sleighs in which supplies (sufficient for three months) for the journey are to be hauled. Regnor Dahl arrived at Eaton Station after dinner; he wanted to hire a man in place of Frederik Larsen, whereupon Wilhelm Kjellmann told him that he had received one man already; if that one was not acceptable—if Frederik Larsen was not good enough—Dahl could not get any better man at the station. Dahl then went to Otto Leinan and wanted to hire him to go along, but since Leinan was in the service of the government, he had to address himself to Kjellmann, to get approval. Leinan was permitted to go, but Dahl had to agree to pay two and a half dollars for each day that Leinan was away from the station.

The government mail came to Eaton Station with one of the corporals (U.S. Army) from St. Michael, and two Eskimo men. Since Kjellmann had important papers to send to the States, he asked the mail run people to wait, to stay over until the next day. This mail is sent out from St. Michael by U.S. Army Captain Walker and is the safest mail to go across Alaska. It is sent out only once each winter, and is separate from the regular monthly post.

## December 7

— The mail run left early in the morning by dog team. Everything has been prepared, ready for the journey to Cape Nome. Kjellmann, Lindeberg, Engineer A. Spring, Alex Joernes, six other Norwegians, and a group of Lapps are going. Kjellmann is leaving Dr. Gambell as manager in his place, and Magnus Kjeldsberg as labor foreman, while Rolf Wiig is to watch over or run the general store and to apportion rations for the people who remain here. The men going to Cape Nome were to drive away across the ice at ten o'clock in the forenoon, all at the same time; but everyone did not get away with such a happy start as hoped, for the untamed reindeer leaped, strained, pulled and tore out in all directions, as soon as they were to pull a sledge behind them. Instead of running along in front of the sledges, they strained backwards. They kept at this until they exhausted themselves and lay down, so that no one could get them to move. All day until sunset the men worked and at last managed to unharness those reindeer that would not go. Then, although it was late, they set out in order to catch up with the others who had gone on ahead; they had to travel all night to reach the others, so that they should not be left behind on the long journey.

## December 8

— Everything went well at Eaton Station. The weather was milder than usual; it snowed a little, but there was no wind. The people were healthy except for the two men who have lain sick all fall. According to the doctor's diagnosis, they were suffering from scurvy.

## December 9

— There is no further labor activity at the station, as the people have to rest for a while. Mild weather with a little snow.

## December 10

— The regular monthly mail team passed Eaton Station at dinnertime. There were four men, two Americans (U.S.) and two Eskimos, with a large dog team for two sledges, which carried also their supplies and sleeping gear. Dr. Southward came downriver from his camp. Since his companion, A. Spring, had gone north with Kjellmann, Dr. Southward is going to stay at Eaton Station for the time being, at least while Spring is away.

## December 11

— Sunday. Dr. Gambell found that some letters had not gone with the mail, so he asked Alfred Hermansen to choose a good sledge reindeer to harness up. Alfred got hold of a sledge reindeer in an instant, hitched it up, and hopped into the sledge himself, whereupon away he went on the river ice, so that one could see the snow spraying behind him. In a way, that the mail was forgotten was Dr. Gambell's fault, but since he was not feeling well, he did not remember that in his office were letters that were to go with the mail. He only realized this the day after the mail run had passed by.

## December 12

— Missionary Karlson from Unalakleet came up to Eaton Station to visit Dr. Gambell, who is sick and in bed. Karlson said that there was news from Golovnin Bay that many people had been traveling to Cape Nome from Council City and that many had had to turn back with frozen hands, faces and feet.

## December 13

— The herdsmen moved the reindeer to the north side of the river, about three miles from the station, into a forested river valley. Here they are to have their winter camp, for there is sufficient food for the animals and shelter from the wind. There is also adequate firewood.

## December 14

— Five men made ready to travel down to St. Michael in order to fetch supplies of food for the station; three are to go with strings of reindeer and sledges, and Paulsen and Kjeldsberg are going especially to take two people—a man and a woman—to Unalakleet for a Christmas visit to the mission there. Dr. Southward left for his cabin to bring down some supplies. For this, he took along a sledge and went up the river on the ice.

## December 15

— Three men with sledges, and the strings of reindeer, drove off to St. Michael. In the evening, Olai Paulsen and Thoralf Kjeldsberg left for Unalakleet, driving (reindeer) sleighs, and they plan to go on to St. Michael in the morning.

## December 16

— The two men who have been lying sick were not improved, but everything else was well. Dr. Gambell was nearly well again.

## December 17

— One knew that it would soon be Christmas. As is the custom, something enjoyable and pleasant is prepared and properly arranged for the Christmas season, but it seems this time that all this would be lacking and that it would be like the other holidays for us here, in that we should spend Christmas Day just like any other day. Dr. Southward came down from his cabin, dragging some supplies on his sledge. He had about twenty-five miles to travel with his load and he traveled all day; so he arrived at Eaton Station late in the evening, tired and exhausted. He went straight to bed, as sweaty as he was. I had just prepared some reindeer meat and some bouillon, and I woke the doctor to invite him to have something to eat. Since Kjellmann left, Dr. Gambell asked me to make the weather recording—what the thermometer showed each morning at 7 o'clock—so that he could register the information in his journal. I am going to avail myself of the opportunity, to make the same notation of temperature in my own journal as well. The weather has been relatively good, so at times the thermometer showed 25 degrees above zero, and sometimes down to 10 degrees below zero. (These values are always in Fahrenheit-thermometer scale.)

## December 18

— Sunday. Some men arrived from someplace up the Unalakleet River where they had been staying all fall. But they had heard so much about Cape Nome that they had broken camp, thrown all of their things on a sledge, and had dragged it themselves (lacking dogs) to Eaton Station. The temperature was 10 degrees below zero.

## December 19

— All that was done at Eaton was to cut and saw enough wood for Christmas. If there was to be nothing else, at least we should be warm. The temperature was 11 degrees below zero. Wind, and clear weather.

## December 20

— Calm and clear weather, temperature 30 degreees below zero.

## December 21

— Calm and clear weather. It was not possible to read the temperature exactly since the mercury was frozen (in our thermometer). This day we had sunshine for three hours.

## December 22

— E. Engelstadt, from Unalakleet, came up to Eaton to look at our new station that was built last fall. Temperature 20 degrees below zero. Calm and clear weather.

## December 23

— The reindeer and sledges returned from St. Michael, the trip having gone well and good. Unfortunately, they could not buy any whiskey, for Dr. Gambell had written to St. Michael that none of the personnel from Eaton Station should be allowed to purchase it. As a result, the Lapps felt unjustly treated and were angry with the new manager at Eaton; for Wilhelm Kjellmann had never forbidden them a drink for their thirst so long as they could pay for it with their own money. Temperature 4 degrees above zero.

## December 24

— It was now the time to welcome the Yuletide holiday—in peace, without any work activity. Olai Paulsen and Thoralf Kjeldsberg, who as previously mentioned had gone on a trip to St. Michael to carry a man and a woman to Unalakleet for a Christmas visit, returned to Eaton Station in the afternoon. The U.S. Army corporal (or officer) from Nulato, with the government mail, arrived in the evening. He will spend the night, intending to drive to Unalakleet in the morning.

At 6 o'clock in the evening came the sorrowful news from the nearby house that Klemet Nilsen had died, after a long illness. He left a wife in Karasjok, Norway, where he was born and had his home; but he himself traveled to Alaska to die.

The temperature was 10 degrees below zero. Wind, clear weather.

Today, December 24th, "No Whiskey" was written, traced out as artwork in the ice on all of the windows of our hall, so that all who entered could see it. (It was done as a joke.)

## December 25

— Sunday. Christmas Day. Christmas morning dawned with good weather. The Lapps put on their finest clothes and began early in the morning, strolling arm in arm, to go from one house to another, visiting for the rest of the day. That is their custom, on Sundays and holidays. Many Eskimos had come around to the station; all who came were given to eat as much as they wanted. I was told that food is always given to all who come to the stations or missions. Thus Christmas Day went for us at Eaton, as on the other holidays, without any special entertainment. But everyone received a piece of reindeer meat, to fry or boil for Christmas dinner. The weather was pleasantly calm and clear, the temperature 9 degrees above zero.

## December 26

— Since it was good weather and we felt lonely so far from our families, several of us lads decided together to set out on a trip to Unalakleet. We looked around there, and went into the schoolhouse where we heard songs and readings (some scriptural), and we watched as the children received their Christmas gifts. The items given were of value proportionate to how good each one was in singing and in reading. In the afternoon we went again to the schoolhouse, where Missionary Karlson delivered a sermon in Swedish language about the birth of Jesus. Then Miss Johnson played the organ, and Engelstadt sang a Swedish song, "God be praised, my soul is saved." The Eskimos had gathered here also, and the schoolhouse was entirely full, of people of all ages. We had intended to leave for home immediately after the program, but all of us were invited to supper over at the mission, so it was late in the evening before we were ready to set out on our walk back. By the time we reached Eaton Station, everyone had already gone to bed, and the last ones of our group did not arrive home before midnight. Good and clear weather. Temperature 9 degrees below zero.

## December 27

— Two men drove down to Unalakleet with five reindeer to get Miss Johnson and an Eskimo girl, Alice. Miss Johnson is a teacher for the children. These ladies had promised to come up to Eaton Station to decorate a Christmas tree, as well as to show some lantern slides, if someone would come for them with reindeer. They arrived after dinnertime and went to Dr. Gambell's quarters. After they had warmed themselves and freshened up, they set about decorating the Christmas tree. That was soon finished, after which everyone received a paper bag full of candies. Then lantern slides were shown, and when that was over, Miss Johnson and Alice sang some English-language songs. After the songs, the entertainment ended.

## December 28

— Dr. Gambell sent two men with six reindeer to Unalakleet to fetch Missionary Karlson and two visitors there from St. Michael. The visitors were Customs Officer Hatch and his wife, who wished to come up and see how things looked at Eaton Station, as well as to see the reindeer herd. They arrived at Eaton in the afternoon. Miss Johnson and Alice had prepared dinner down on the first floor of our building, in Wilhelm Kjellmann's quarters, where all of the rooms were warm, and where they are residing. A little before sunset, the herdsmen came, driving the herd past the station. Our visitors were there and viewed the herd, which evidently was a curiosity for them, since they had never seen so many reindeer in one group.

## December 29

— Dr. Gambell was up early in the morning, made a fire in the stove downstairs, and then set about making breakfast. I had already prepared breakfast upstairs, and since Karlson, J. Brynteson and Hagelin were up in our hall, we invited them to join us. They were already at table when Dr. Gambell called them to come downstairs. Most of the guests went back to Unalakleet in the afternoon, but Hatch and his wife are to remain at the station until the next day since there is a storm and driving snow. Temperature 20 degrees below zero.

## December 30

— Alfred Hermansen drove the rest of the guests down to Unalakleet; no others here are so able as he is in this, for he knows well how to handle the reindeer.

A grave was dug for the deceased Klemet Nilsen. Kjellmann had designated a burial area on a small, round sand hill, to be used as a cemetery.

The weather was stormy, with snow. Temperature 5 degrees below zero.

## December 31

— We all have in our minds that this is the last day of the old year, and that we should soon enter into the new year and forget about all of the bad things we had experienced in the old year. We should think just of the future, and that we all shall receive only everything that is good.

The total number of reindeer slaughtered between fall and Christmas was about seventy, of which the greater part had broken legs and to slaughter them was necessary. Some others had diseases that made them useless for anything, and they, too, had to be killed. One might easily conclude that the reason so many reindeer have to be destroyed following injury to their legs is that the Siberian animals do not have nearly so strong a bone structure as do the Norwegian deer. The Siberian reindeer have insufficient bone quantity in that the bone layer appears thinner, and the bones are filled with marrow and blood, so that in thickness (diameter) the bones are approximately half that of the Norwegian reindeer. So the bones of the Norwegian reindeer are hard, and those of the Siberian, softer; after a bone from a Siberian deer has been cooked (boiled), one can chew it to pieces. The Lapps maintain that this comes from the fact that the Siberian reindeer always are walking on ground that is soft, while the reindeer in Norway have a lot of stone and rock to walk upon.

This past year, my journal has been recorded daily. The journal for 1899 will be only by the week.

Temperature today 0 degrees.

# Weekly journal for 1899. Eaton Station and Cape Nome.

## January 1–January 7

— 1st Week. Missionary Karlson from Unalakleet was in Eaton Station to hold a service. Wilhelm Basi translated his sermon from Swedish into Finnish, so that the Lapps could understand it. Karlson delivered a funeral sermon for the deceased Klemet Nilsen, as well as the committal service at his grave.

On Monday, 2 January, nine men set out for St. Michael, stopping overnight at Unalakleet; from there, five men from Unalakleet were to accompany them. They took along thirty reindeer, which were to bring back loads to the station. Some of the people who remain at the station were ordered to saw wood, and others are to cut firewood and carry it to the houses.

A Lapp woman gave birth to a daughter.

Peder Berg also went with the others on the trip to St. Michael, to undergo a medical examination, since he has had pains in one foot, for which Dr. Gambell could do nothing. Peder had to return because his driving reindeer had balked and would not travel with the company. The animal was named Moses, and he was in the habit of jumping on his driver whenever he felt like doing so. In that respect, Moses had not neglected his compliments now, either. Instead of running along after the other reindeer, he began to ram and kick Peder each time Peder tried to get Moses going. All his efforts were in vain, and so Peder Berg had to turn back, and he reached Eaton in the evening the day after they had left the station. Peder said that Moses had given him more hard blows than he could count—sometimes with his head and sometimes with his hooves—and Peder's jacket was torn to pieces, and his foot was worse than it had been before.

Dr. Gambell and Dr. Southward went down to Unalakleet with a measuring line, to determine how many miles the distance is from Eaton to Unalakleet, in a straight line. They returned in the evening, having measured the distance on the trip downriver, and they found that it is seven and three-quarters miles, from Eaton to Unalakleet. While they were in Unalakleet, Dr. Gambell received a letter from Wilhelm Kjellmann, in which Kjellmann reported that he lay ill in an Eskimo cabin about fifteen miles north of Golovin (at Golovnin Bay) and wished for medical assistance if that were possible. Missionary Anderson had visited him and had given him some medicine, but he was not well. Some men were sent at once to the herd, to select three of the best lead reindeer and bring them down to the station, in order to depart the next morning with Dr. Southward; Dr. Gambell would remain at the station. During this week, the mail passed Eaton Station on the way north to the Yukon River, with two men, five dogs, and a sledge. All week we had mild weather with wind. The temperature varied between 10 and 4 above zero.

## January 8–January 14

— Ole Baer and Dr. Southward drove away on Sunday at midday. Each of them had his own driving reindeer, and there was one to carry the supplies. Dr. Southward had on a complete Lapp winter outfit. This tall, full-grown man was wearing reindeer-leg trousers, a coat with wide belt and a large sheath and knife hanging from the waist, and a Lapp cap and gloves, so that he appeared much taller than he actually was. And furthermore, this was the first time he had been dressed in such clothing, and it made him stand as straight as a poker. He had to lie down on the sledge when the reindeer began to move, which was also new for the doctor. Ole Baer had to take him in tow, as well as the third reindeer. And thus they set out, down the river ice, so that one could see the snow spraying behind them. They were not going to stop in Unalakleet for the night, but would drive as fast as possible in order to arrive at their destination as quickly as possible.

On Thursday, Dr. Gambell went down to Unalakleet. In the afternoon, the people returned from St. Michael. Dr. Kittilsen, who had been on a trip north to Golovnin Bay, returned as well.

In the evening at sunset, Isak Nakkila returned from his unfortunate hunting trip; he had been about three miles from the station, to shoot ptarmigan. While he was walking, he happened to fall and twist about in such an unfavorable manner that the gun went off. The shot struck his left arm, and the whole load went through his elbow, leaving a lot of buckshot in his left thigh, above the knee. As he himself told us, it happened at dinnertime (about midday), but he had lain unconscious for a long time. Word was sent immediately to Dr. Gambell, who arrived within a short time. Dr. Kittilsen also was present when the wound was dressed. Although it was 25 below zero that day, Isak had not frozen himself.

Dr. Gambell dispatched two men with some reindeer to St. Michael to get some supplies for the North American Transportation and Trading Company, which were to be transported to Cape Nome. Another group of men with reindeer were to set out on the trip to Cape Nome at once, with supplies, and also to deliver a report to Wilhelm Kjellmann regarding the reindeer that the Mission at Anvik was to receive on loan from the government. But inasmuch as the mission at Golovnin Bay had impounded all of the animals that were there, the prospects are that Anvik would not get any this winter—especially, too, since the Cape Nome traffic was so heavy that all the reindeer were in use. Nevertheless, Dr. Gambell wanted to report this to Kjellmann, whether or not he had any reindeer for Anvik.

From these travelers going between St. Michael and Cape Nome came news from the gold fields, where many had begun to jump others' claims, thereby acting fraudulently. An appeal had gone to Captain Walker (military) in St. Michael, for soldiers to be sent to keep the peace. Otherwise, it seemed that the disturbances up there would lead to violence and distress.

The weather was calm and clear, temperatures between 30 to 37 degrees below zero.

## January 15–January 21

— Two boys from St. Michael, who were on their way to Cape Nome, came to visit Eaton Station on Sunday afternoon. They said that Captain Walker was going to send some soldiers north, as the people had requested.

On Monday, two men went to St. Michael, with some reindeer. On Tuesday, at midday, Magnus Kjeldsberg and Dr. Kittilsen departed from Eaton for Cape Nome. They had with them five driving reindeer. Those gold prospectors who had built winter cabins for themselves up along the Unalakleet River, upstream from Eaton, came down to the station, each pulling his own sledge full of supplies and clothing. They were on their way to Cape Nome. There are four of them, and they are the last of those who were living upstream by the river.

At the beginning of the week, it snowed for three days, but later it was clear and calm, so all week was cold. Temp. between 10 and 25 degrees below zero.

## January 22–January 28

— Illness had set in, and once again two Lapps are lying ill with scurvy. Many others were ill, with various afflictions.

The men who had gone to St. Michael the previous week returned on Wednesday. Mr. Hvidsted accompanied them to Eaton. He worked for the North American Transportation and Trading Company in St. Michael as a store clerk, but he had a claim at Cape Nome and now, due to the disturbances at the gold fields, he was going north to look after his interests. Since arrangements had been made for a group of people to travel northward on January 30th, and since transportation was available, Mr. Hvidsted would travel with that group.

A snowstorm has raged all week, and the extreme cold has also been bitter. Temp. between 20 and 30 degrees below zero.

## January 29–February 4

— We received tidings from Cape Nome once again on Sunday afternoon, when a man by the name of Haster came to Eaton Station. He had a cabin above (north of) the station, and there he had his supplies. He and another were in partnership, and his companion had stayed behind at Cape Nome. Haster said that only three claims had been jumped, as far as he knew, two of them belonging to Olai Paulsen and one belonging to Magnus Kjeldsberg. He had heard this at Cape Nome through a report. Those of our comrades who had gone with Wilhelm Kjellmann also reported the same to us by letter. Our comrades said as well that all were in good health, that they were short of supplies, and also that Wilhelm Kjellmann had recovered sufficiently so that he had continued his journey farther westward and he was now among his men at Cape Nome. Dr. Southward had told those of us here at the station that Kjellmann would have to travel down to the United States during the summer, to undergo a surgical operation, for otherwise he should not become well completely.

From Cape Nome on Tuesday, January 31st, we received reliable news from the gold fields when two men from our station who had gone with Wilhelm Kjellmann returned and reported that everything was fine, except that Kjellmann was not completely well. All those who had gone to the gold fields had staked several claims for themselves. Town lots were also laid out along the seashore, at the mouth of the Snake River, which Kjellmann and Engineer A. Spring had staked as placer mining claims. The fellows then had the opportunity to acquire town lots for themselves on the west side of the Snake River, and a contract subsequently was written to the effect that they had purchased each lot from Kjellmann and Spring for $100. The lots were 300 feet long and 100 feet wide. No lots were sold on the eastern side of the Snake River, where they thought a town would be built (and where in fact it was being built). Engineer A. Spring had worked all winter, measuring and laying out the town, with the streets and all in a standard and uniform pattern. The actual construction of the town and the selling of lots presumably would take place in the spring, when (again presumably) people would be streaming in. We heard a little about the conditions with relation to an accident that had occurred: A young man had frozen completely both of his legs. He had been with his companion, up in the mountains to stake gold claims, and had fallen into a river. There was no wood for a fire, there was only cold and storm, and so his comrade had him climb into a sleeping bag; then his comrade ran down the mountain. He came to Wilhelm Kjellmann and told him of the situation. Kjellmann immediately sent out Ole Olsen, a young Lapp, with reindeer to bring the fellow down. It did not take Ole long to get there. He first gathered up a little of the wood of the kind that is to be found there and made a fire, so that the frozen fellow could warm up a bit. Then he placed him on the sledge and drove him down. Dr. Southward took care of him and told him that his legs would have to be amputated before they became gangrenous. A letter then came to Dr. Gambell at the station, from Dr. Southward, asking that surgical instruments and medicine be sent at the earliest opportunity.

At Cape Nome, Dr. Southward was employed by Dr. Kittilsen (who had the position of recorder) with wages of $10 per day.

The weather was unstable. We had snowfall nearly the entire day of January 31st, until the snow was four or five feet deep. It has been stormy, but still the temperatures were between 18 and 27 degrees above zero.

## February 5–February 11

— On Monday, Dr. Gambell had three men drive to St. Michael with eighteen reindeer, to pick up some soldiers under authority of the military captain and convey them north to Golovnin Bay, or also to Cape Nome, as this had been requested for the purpose of establishing order. On Wednesday we received more news from Cape Nome, when two of Wilhelm Kjellmann's crew arrived. They had to leave the gold fields because the supplies ran short. All of them wanted to return in spite of that, but Kjellmann would not allow it. Kjellmann had wanted three houses to be built, and they had remained in order to put up the last one. In the construction, they had taken their materials from the shore, where logs had been cast by action of the sea. They had dug the logs out of the snow and had sawed them up and built houses. They told us that while they were there, many people had been at Cape Nome staking claims, and to spend the winter had then gone back to Golovnin Bay, St. Michael, and all about to other areas where supplies and housing were available. Among these people, there was never a mention of what the costs might be, only of the fact of securing supplies and housing. In this connection, I shall mention the merchant John Dexter in Golovin; he aided all travelers who needed food and lodging. He also has given supplies on credit to many who had no money with which to buy anything, and he won the praise of the people. He indeed has merited that. I must also not forget Missionary A. Karlson and his people, and to say here how hospitable they are, and how they have been aiding many who needed help.

Everything at Eaton was quiet as usual since last week, because of the storms. It was not possible to work further, and the last week a strong wind blew with driving snow; but the temperature was 20 above zero, later changing to 8 below zero.

## February 12–February 18

— Everything was well and good at the station, but scurvy was making itself felt. Many of the Lapps, including the women, were hit by it, and many of the men were in bed. Dr. Gambell had a constant struggle with them, for they never listened to what the doctor told them; nor would they use the medicines that the doctor prescribed. They wanted to have other medicines, or something like whiskey, which stung the throat; or something for the nose, which would cause the eyes to water. They thought that such would cure them, but not such medicines as the doctor prescribed, which have neither taste nor smell. Although the doctor told them that they had scurvy, they can stand and deny that it is scurvy—though it would seem that they actually had symptoms of the disease.

On Wednesday, another of Kjellmann's crew returned to Eaton, as Kjellmann had ordered him to accompany Missionary Karlson, inasmuch as there was room for one man on the reindeer sledges. He reported that some others, including Kjellmann, would return soon to the station, but that a number of those who had earlier traveled north with Kjellmann were not coming back. They were going to remain at Cape Nome for the rest of the winter; they were no longer in the government's service but were their own masters. Many of us at the station were waiting for Kjellmann's return, thinking that we could travel north to the gold fields, and leave his service when our contracts were terminated.

All week we had clear weather, but it was very cold. The temperatures were between 25 and 40 degrees below zero. I thought about old Norway—about how even at this time of year it was not so cold there as where I now was. Here, when there is wind along with that cold, it is nearly impossible to be outside. Even the Eskimos, who are so hardy and used to the Alaska winter, came running into the kitchen many times to warm their noses. But I also saw an Eskimo woman who sat outside in the snow, loosened her clothing, and nursed her infant at 30 to 40 degrees of cold, in wind and driving snow. Missionary Karlson had this to say about Alaska's coldness one winter day in February: "O, you cold north, merciless and unloving! You drive your children away, or you kill them. You laugh, but your laughter is only my judgement on your lips. If only you had been milder, if you treated me more gently!" So we can understand that Karlson, who has lived here about thirteen or fourteen years, has seen many of Alaska's winter evenings, and that not all of them have been the best. First he came alone amongst the wild people, and next to a deserted and out-of-the-way place where there was no contact or communication with his own people—aside from St. Michael—and there to begin his life's work in a sod hut (perhaps it was of wood, I cannot recall). But he came alone to Unalakleet, where he was able to put up a cabin, with the help of the native people, in which to spend the winter. The following summer he was able to improve the cabin a little, and thus year after year, whenever he had time and opportunity.

## February 19–February 25

— As there was nothing to do during the idle and lonely hours, two Norwegian fellows, about 21–22 years of age, fell in love, each with his own Lapp girl. They are to enter into matrimony the following Sunday, and the weddings are to be held at Eaton Station. So now there was more life than usual at the station, with arranging and planning everything for a joyful celebration.

With the soldiers, the men arrived from St. Michael on Saturday evening at 12 o'clock (midnight); so the entire night passed without rest, as I was the cook, and had to get up and prepare food for those who came. It was morning before everything had quieted down. There were five soldiers, a corporal, and a lieutenant who were going to Golovnin Bay. The judge from St. Michael was going also, traveling by dog team; he waited at Unalakleet while the soldiers came up to Eaton to spend Sunday. The transport animals and drivers were to be changed here at the station. Olai Paulsen, Ole Stensfjeld and Isak Tornensis are to take the soldiers to Golovin and Cape Nome. Olai and Ole were not going to come back, but rather would stay at Cape Nome for the rest of the winter.

There was clear and cold weather all week, and temperatures 30 to 36 degrees below zero.

## February 26–March 4

— Everyone was up early on Sunday morning, as planned, and at 9 o'clock the wedding parties drove down to Unalakleet to be married by Pastor Karlson. Each wedding couple had a sledge where bride and groom sat, drawn by two good sledge reindeer, while the rest of the wedding party rode independently. While they were at Unalakleet, I was to prepare the food for the wedding party as well as for the soldiers, and have everything ready when they returned from Unalakleet. At 5 o'clock in the afternoon we heard the sound of bells. In half an hour the wedding party was back at the station and had gone to the main hall. The bridegrooms and their brides sat down at the prepared table, along with others of the wedding party, as many as were places at the table, to have a cup of hot chocolate. The wedding couples sat side by side, each groom by his bride. One couple was dressed in ordinary peasant clothing, while the other was in full Lapp costume. That bride was adorned and dressed in the best fashion, with silk and special cloth. On her head she wore a cap of the typical Lapp style (for women). On this was placed a crown, on which were stars of silver and other decorations, and from which at the back a long, silken ribbon trailed. On her feet she wore gray Lapp moccasins, fine and lovely ones, sewn from reindeer skin. The bridegroom himself looked like an officer in parade dress, with a broad, silver-inlaid belt about his waist, and about his neck a broad, white band that crossed his chest diagonally and fastened beneath the belt, so that the fringes at each end hung far below his tunic. He wore on his head a Lapp cap with the traditional four peaks and on his feet snow-white Lapp moccasins. Next the bridegrooms went about with the flask and poured for everyone. The bottle went around slowly, until someone began the joik—the rhythmic, monotonal chant of the Lapps. In such a ceremony, by custom when the chanting begins, the Lapp woman, as she has become a bit intoxicated, throws herself affectionately on the men of the party. It can be anyone at all in the party, so long as he is a Lapp.

Fresh reindeer meat, the best at the station, and bouillon, had been prepared for the supper, which everyone could enjoy. When supper was finished, the glasses were busy again. Both wedding couples were present at 11 o'clock in the evening, when the wedding was to be over. At this time, the grooms poured another round for everyone present, after which the celebration was to conclude. They gave money for bridal gifts, as is the custom, and then all went to their houses and all became quiet at the station. We heard the next morning that on their way home, to the nearby houses, from the main building, two of the Lapps had fallen or rolled into the ditch by the road, with their wives; they were all intoxicated, but the women had been injured. One was with child, and she was now gravely ill; the other had been made unconscious but she had come to her senses, though she was still in bed. The wives said that their own husbands had caused this—pushing them into the ditch after they had made some remark, accusing the men of something.

On Tuesday Thoralf Kjeldsberg and Wilhelm Basi left Eaton Station, not intending to return. They had managed to get hold of five dogs as their transportation animals, as there were no reindeer available for them. In addition, freight was so expensive that it was not possible to obtain reindeer freight. The cost was thirty to forty-five cents per pound, for delivery of freight over to Cape Nome. The two drove away with these five dogs, of which only one was accustomed to being a sledge dog. The fellows had loaded the sledge with over 300 pounds, but the dogs easily pulled it over the ice and on the good snow. Thus they left the station. They had bought supplies of flour, pork, coffee and sugar, and they had to pay as high a price for these as the Eskimos must pay; but that was of little concern to them as long as they were able to obtain what they needed and got away.

On Wednesday it was the soldiers' turn to depart on a journey. They took twenty-six reindeer and three men from Eaton Station. As previously mentioned, Olai Paulsen, Ole Stensfjeld and Isak Tornensis became drivers for the strings of sledges. There are two new patients with scurvy, so that now seven are ill. Dr. Gambell had quite a task with these contrary patients who never heeded his advice, nor did they bother to take the medicines that were prescribed for them.

The weather was clear, but it has been very cold all week. The temperature has stood constantly at 30 to 38 degrees below, except for Saturday when it was 16 degrees below zero. There has been a wind, which bit the nose so bitterly that I must say it as Missionary Karlson put it once to me, when he felt that the cold would hurt his nose while he was working outside: "My nose bears witness to the fact that it has been cold recently. I believe that the noseless would survive here better than those who have noses." And it is indeed difficult to save one's nose in Alaska during the winter, when there is frost and wind. On Saturday Dr. Gambell was called down to Unalakleet to tend to a sick Eskimo woman.

## March 5–March 11

— Dr. Gambell came from Unalakleet on Sunday for some medicines and returned there immediately. One Lapp woman who had gone into the ditch on the evening of the wedding was very poorly, but the other was already quite well. Dr. Gambell came back from Unalakleet on Monday and said that the Eskimo woman was beyond recovery, for no medicine could help her. The doctor had to take immediate care of the Lapp woman who lay ill. One man returned from the group that had gone north five or six weeks ago with some supplies for one of the companies in St. Michael. His two companions were behind on the trail with their reindeer and they would arrive in two days; since he was driving a reindeer without a sledge, he was able to travel much faster than they. He said that probably Kjellmann also would be at the station within one or two weeks. Otto Leinan came home on Wednesday from his trip to Cape Nome with Regnor Dahl, who had stayed at the cape. Otto came home now only to move his family from Eaton to Unalakleet, where he had rented a cabin in which his wife and children would be able to live during the rest of the winter, until the spring. He was going back to the gold fields again and intended to remain there until the sea was free of ice, and then he would take his family over to Cape Nome. Leinan and Dahl had located numerous claims for the company that had sent them out to Cape Nome. Dahl had staked eight claims; Leinan, six claims, and for Ole Krogh were staked two claims through powers of attorney; and two were staked for Nils Bals because they had his son along to tend the reindeer, since neither Dahl nor Leinan was accustomed to doing that. Nearly all of the claims they had located were on the Nome River, which empties into the sea east of the Snake River about three or four miles from Nome settlement.

Some of those who lay sick here were a bit worse, but some of them had improved. Johan Petter Johannesen from Kvalsund was giving some help as a doctor. He prepared medicine as needed for those who were sick. He bled some, and gave others a rub, and visited his patients two or three times a day. The patients maintained that Johan Petter's remedies really helped, and some were recovering. Many paid him for his attentions and for the inconvenience they had caused him. Johan Petter left the station a few weeks later, having earned about 100 dollars in his work caring for the sick.

Temperatures were 20 to 26 below zero.

## March 12–March 18

— I went down on a brief trip to Unalakleet, where some people who had just arrived from Cape Nome told me that Kjellmann would be back within a few days.

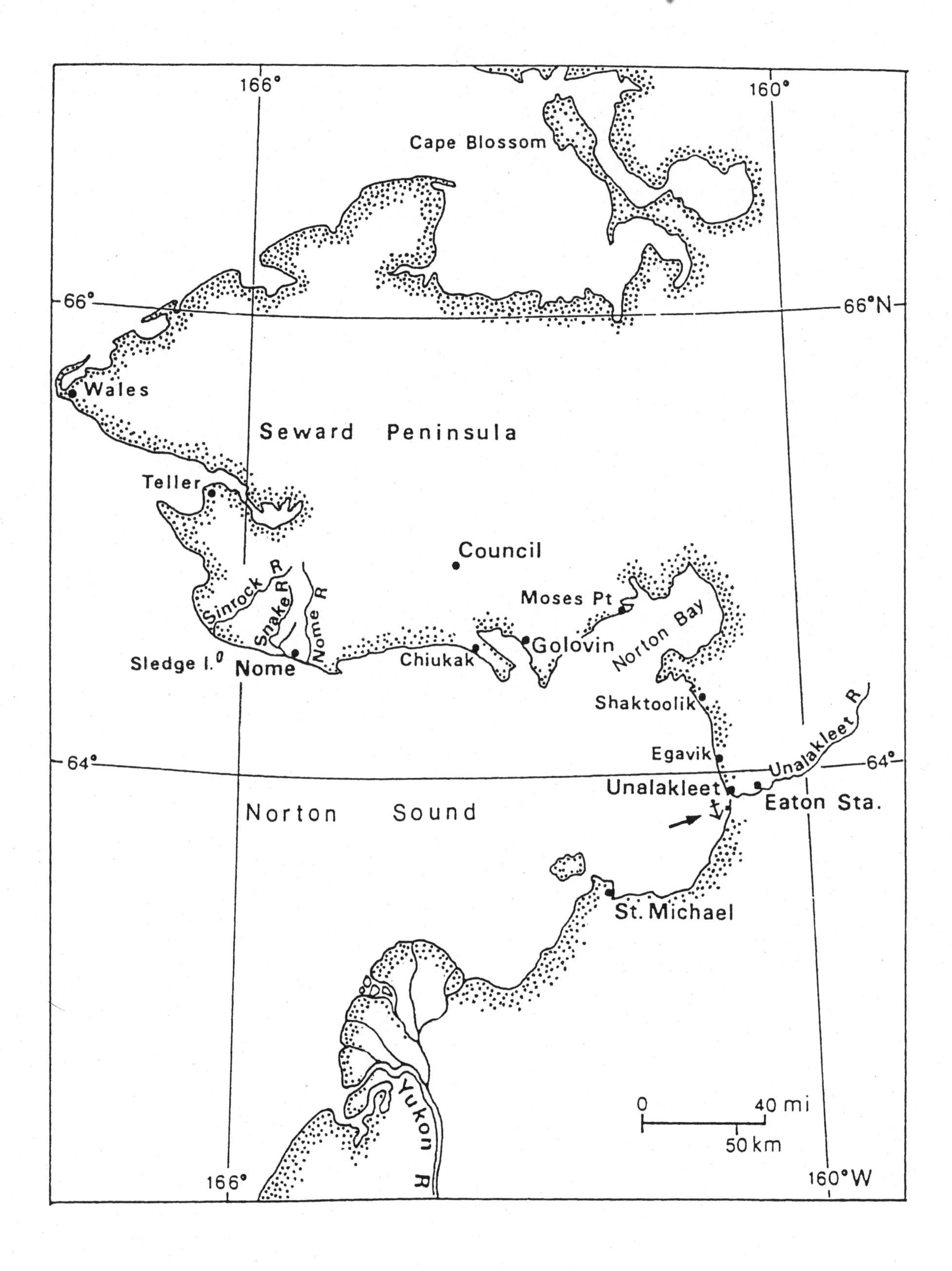
166°
160°
Cape Blossom
66°
66°N
Wales
Seward Peninsula
Teller
Council
Sinrock R
Snake R
Nome R
Moses Pt
Golovin
Norton Bay
Sledge I.
Nome
Chiukak
Shaktoolik
Egavik
64°
64°
Unalakleet R
Unalakleet
Eaton Sta.
Norton Sound
St. Michael
0
40 mi
50 km
Yukon R
166°
160°W

On Tuesday evening, Fredrik Larsen came driving up to the station with a dog team from Port Clarence, where he had stayed for a while with his wife's parents, who live there. Now he was on the way to St. Michael to fetch some supplies that he had left there last summer. He told us that Kjellmann would arrive the following evening. And as he said, Kjellmann arrived on Wednesday evening, together with A. Spring. The others in the company had been delayed and they did not reach the station until Thursday. They all looked very dirty, and nearly all had frozen their noses and cheeks. Newly grown skin was in light and dark patches on their faces, and they had a terrible appearance because of that. They were all well and lively, although they had had a difficult trip. They had been fortunate enough to make many gold claims, and the joy over this made them forget all the hardships they had suffered. They were going to go back to Cape Nome in a few weeks, no longer intending to remain in service at Eaton Station. Three of their companions—Ole Berg, Otto Greiner and Amund Hansen—had stayed at Cape Nome and they would not return to Eaton Station. Mostly, the fellows complained that they had received too small a ration on the trip, and all during the time that they were at Cape Nome, although they had worked hard at gathering up logs, and sawing them and building houses, as Kjellmann had ordered. But even at the station people were complaining constantly, moaning that they had been so stupid as to depart from their poor Norway, where one got to eat one's fill of food now and then, while in the service of rich Uncle Sam, one never got anything but the confounded pork and beans all the time and that it seemed that one could recognize the smell of pork and beans from several miles away. In fact, one never received more butter for a month's ration than one could put into a pipe. It seemed to us that the many illnesses at the station must partly be caused by that; some could not walk, and one had met an untimely death. That people wanted to leave the station to work for private individuals was therefore no surprise, for though the wages were no better, the food was. The Lapps were not concerned about wages, so the manager had granted the wishes of many to go and work at another place that would suit them better. All week we had good weather, with the temperature 10 to 30 degrees above zero.

Opposite: The area of Eaton Station and Anvil City (Nome). The site of anchorage (indicated by arrow) of the *Del Norte* on 31 July 1898 was just to the south of Unalakleet. Carl Johan and Johan Petter walked from Eaton Station to Cape Nome in 1899. Locations of villages through which they passed en route are shown.

## March 19–March 25

— On Monday Kjellmann ordered six men to go about five miles up the river from the station to saw planks for boats. On Friday (Good Friday) Engineer Spring went to his cabin upriver, where he had some supplies and clothing. He carried along a letter

Kjellmann sent to the men who were sawing the planks; Kjellmann wanted them to help Spring in getting his supplies to Eaton. Spring reached them about at mid-day, and the men ceased their work with the saws and went along to Spring's cabin, where they spent the night. On Saturday morning they began the trip down, carrying loads, and did not reach the station until the evening. They were tired and very angry that Kjellmann would order them to haul Spring's supplies when they had been assigned the heavy work of chopping and sawing wood for planks; and also, there were people and reindeer available at the station, who should have been more suitable a choice of laborers to walk twenty-five miles pulling loads than men who already were at work. This week Otto Leinan moved to Unalakleet with his family, therewith leaving service. It seems that all of the healthy younger people who still remain at the station were ready to leave at a moment's notice, to head north to Cape Nome. Everyone was asking Kjellmann to help them out with supplies and a little money; they would make sledges and pull them, themselves. But Kjellmann refused the pleas of many of the men, because supplies at the station were running short. Sale of food items had been open for the Eskimos all winter, and goods worth several thousands of dollars had been sold at the store. And now, people who belonged to the station as staff had to go without and had to postpone their trips.

A Lapp woman gave birth to a baby girl. Tanejja Donnak's wife also had a baby girl. Donnak is an Eskimo who has worked for the service, with salary from the government, for several years, learning reindeer management.

The weather was relatively good. This week we had rain for the first time since last fall. It snowed for a few days as well. The temperature was quite mild, the mercury varying for several days over 32 degrees above zero.

## March 26–April 1

— Our Easter was quiet, as usual for a Sunday, so that no one was obliged to work. We talked about Easter at Haines in southern Alaska in '98.

On Monday, several of the fellows began to build sledges and to plan, gradually to prepare for the journey to Cape Nome.

On Wednesday, John Losvar's wife gave birth to a daughter. That was the fourth birth, and all were baby girls. All week people have passed through, on the way to the gold fields, coming from the Yukon, from Dawson City, and had set out on the long trip after hearing news of gold at Cape Nome. Many who did not have enough dogs to pull their supply sledges were themselves helping the dogs at pulling. On his way to Unalakleet on his return trip from the north, the judge from St. Michael passed by at the station at the end of the week. We heard that two men had jumped the claims of Johan Sp. Tornensis and Mikkel Nakkila at Cape Nome; when the two would not give up in their illegal actions, they were seized by the military and were to be taken to St. Michael for trial. Johan and Mikkel also had to be present.

Good weather, the temperature at times 35 degrees above zero.

## April 2–April 8

— There were eighteen of us men who were ready for the trip, but since not all had sufficient supplies, five were to go to St. Michael to work there until after the sea opened, free of ice, then to go to Cape Nome by steamship. I had been serving as one of the cooks, and I resigned from the service on the 8th of April. We set out on our journey the same day. We were three: Johan Petter Johannesen, Alfred Nilima and I. We had all of our gear on one sledge; it weighed about 800 pounds. Thus we set out down the river, and waved farewell to Eaton Station. The other fellows were to begin their journeys a few days later since all were not yet ready to go; Ole Krogh, Otto Leinan and Johannes Rauna were to go with a second sledge, the third sledge party included John Losvar, Ole Rapp and Lauritz Stefansen, and Hilmar Hansen, Lauritz Larsen and Rolf Wiig would be the fourth party. We arrived at Unalakleet at about the same time as the strings of reindeer sledges, driven by men who were also going to Cape Nome, carrying supplies for Kjellmann. Donnak, who had also resigned from the service, was there, traveling home to Port Clarence, where he plans to begin raising reindeer on his own. Karlson kindly gave us lodging, and with him we spent the night. The weather was mild, with driving snow.

## April 9–April 15, the fifteenth week of 1899

— Sunday morning we loaded our gear, some obtained at Unalakleet, on the sledge, but there was too much and we could not fit all of it on the one sledge and we had to leave some behind. When we were at last ready to depart, a crowd of Eskimos gathered around us, or rather around our sledge, which held now a huge load, arching up to the same height as we were ourselves. We were planning to pull this sledge northward to Cape Nome, about 210 miles—a good distance. But we did not mind that; we intended to pull it as far as we could, as far as we could find strength. Slowly we began to move off; we hoped to accompany the reindeer sledges and their strings of reindeer if we were able. By departing when we did, we should get ahead of them; we passed by them when they were still in Unalakleet. At sunset we reached Egavik, a little village of the Eskimos, about fifteen miles from Unalakleet. We were tired and hungry and did not have the strength to go farther; the reindeer sledges passed by us and as they were planning to go on for several miles, we were left behind.

An old Eskimo man greeted us and invited us to a cabin. The people carried our sleeping gear, our coffee pot, and our food sack, and we had only to walk up empty handed, following as they led us to a little wooden hut, where we were to enter and spend the night. An old man lived alone in the hut, so there was plenty of room inside. We boiled water for coffee on the hearth (there was no stove, only the hearth in one corner), fried some pancakes and a little pork for supper, and shared these dishes with the master of this house. Then we crept to a corner of the cabin, and into our sleeping bags and soon we all had fallen into peaceful slumber. Early in the morning, our host was up and made a fire on the hearth, where we made coffee and fried pancakes. We set out again early, before the sun rose, as we hoped to catch up to the reindeer sledges, which we did, but they were ready to leave just as we reached their camp, while we had to catch our breath, so we were left behind. We rested for a while, then we took hold of our sledge and actually pulled it all day, with the thought of reaching the Eskimo village of Shaktoolik by evening; but night fell and we had nearly given up hope of arriving as hoped. But then we heard dogs baying, so we assumed that it could not be so very far to the village, and with all our strength we set about getting there. But the sledge was heavy and since there was much frost, conditions were not good for the iron sledge runners; that was the main problem. Nevertheless, we toiled forward, nearly to the houses. We could see people standing on their huts (on the roofs, which the Eskimos customarily use when they want to look around with better view of an area). We threw ourselves into the snow in order to cool ourselves off; we could not manage to walk any farther. Before we knew what was happening, a whole group of the people came running up, people of all ages, women and men. Using sign language, they indicated to us that they wanted to help us by pulling our sledge to the huts. We let them do this, while we only had to walk behind the sledge. They led us to one of the underground houses, and we took our sleeping bags and the food sack and went in, crawling first through a little hole into a dark passageway which had footholds and angled downward to another dark hole. Here we fell three or four feet deeper, down into a narrow tunnel that led into the room where they lived, into which we entered, again through an opening; we had to crawl on all fours to get in. In the room, many Eskimos sat around the earthen wall, women and men, half naked, and we sat down in the middle of the room. The women at once took our coffee pot, filled it with water, and placed it on their fire. They passed to us a large, oblong, wooden bowl-like tray of cooked seal meat and indicated that we were invited to eat, so we took the tray on our knees and cut generous pieces of meat for ourselves. We gladly ate, for we were hungry, and it did not matter that the tray did not seem to be entirely clean. We were looking at the seal meat and not at the tray. It was well cooked, and it tasted good to all of us. We had a cup of coffee, and felt satisfied and grateful. The people themselves ate some frozen lingonberries with seal oil, and drank tea for supper. Later on, I observed how they washed the children with urine, and stripped hides in it, and washed and soaked everything in it. I thought that all was not going to go well for me, when I thought about the seal meat I had eaten not so long before. Roundabout all over were containers full of urine. They evidently never threw it out, so quite a vile odor came from this and from the seal oil, which stood inside in wooden containers, turning sour and appearing rotten. We could not see what in the room might be soiled or what might merely be black, as it was dark inside; a single light was in one corner, and that and the coals on the hearth were all they had for lighting. When it was time to go to bed, we placed our sleeping bags on the earthen floor and crept well inside them so that we should not be upset by the odors. When we lay down, the light was put out and it became dark as the grave. But we fell asleep and awoke in the morning after a peaceful night. After breakfast, we went outside to see if we had good traveling weather. We saw the people with the sledges and reindeer, camping about half a mile from us, by the seashore. We had passed them during the last evening, in the dark, and now we were ahead of them. I put on my snowshoes and went over to them, to ask whether they planned to travel today, since the weather was bad, with a storm and snow, and poor snow conditions. But they were going to leave anyway, in order to get across the isthmus to Norton Bay, where they were intending to let the reindeer rest a bit. Then, early on the next morning, they planned to cross the fjord, which was a whole day's journey on the ice, about twenty-one miles. I returned to my comrades and told them that the reindeer sledges were going to go, whereupon we packed our gear and thanked our friends and set out in order to stay ahead of the reindeer. We had not gone very far before we became very tired, since the snow conditions were so poor. We had to stop at an Eskimo house, where a single family was living. They kindly asked us to come inside, and we stayed there the rest of the day and overnight. The next morning, since the weather was better, we set out again and in the afternoon arrived at the shore of Norton Bay, where we put up our tent. The worst thing was that there was no firewood to be found. But without wood we could not operate a stove on the ice anyway, as the fjord was too wide to cross in one day and we were going to spend the night on the sea ice, so we decided to settle down to wait until the following day. Since the reindeer sledges, which had passed us yesterday, evidently had crossed the sea ice before we reached the shore of Norton Bay, we were now far behind. Several with dog teams and with reindeer sledges arrived in the evening and set up their camps there on the beach where we were—including Johan Tornensis, Inge Bals, Isak Haetta and Ole Klemetsen, all of them on their way to their claims at Cape Nome; Inge had been hired by Dr. Kittilsen to cook for him at Cape Nome for the coming summer. Johan Brynteson and J. Hagelin were also with the group, headed north to their claims, but since they were driving reindeer, they would certainly have an easier trip. In the morning we all began to get ready for the journey. Johan Tornensis kindly took about 200 pounds of our load on his reindeer, which he would transport across the ice to the other side of the fjord, so now we pulled our sledge easily. When we reached the middle of the fjord, a good wind came up, and we made a sail out of our tent. This was a great help, since the snow conditions were so poor. We came to the other shore at 7 o'clock in the evening, tired and hungry, for we had not taken time to eat during the day, hurrying on with all our strength in order to reach land before dark. As soon as we got off the ice, we set up our tent. Not far off, in a little sod hut at the foot of the mountain lived an Eskimo family; they had recently been seal hunting and had fresh seal meat. We bought a piece, cooked it, and ate it for supper. Afterwards we crept into our sleeping bags and slept until after midnight, when it became so cold that we all had to get up and move around. We built a fire, made coffee and arranged our things as best we could then, all in good spirits and hoping that at Cape Nome we might find reward for all our toil and labor on the journey. We set out again and reached some Eskimo houses at about midday. We presumed that Johan Tornensis had left our supplies here, while he and the others had gone on farther, and we found that to be the case; as soon as we came to the houses, the Eskimos pointed to the sacks that Johan had placed on one of their platforms—the sacks were ours. We inquired about how far they thought the reindeer sledges had gone, whereupon they showed us that an Eskimo sod house was located on the shore, about four miles farther westward, and they said that the "lav Lakker" (Lapps) were at the house and had set up their tent there, so we threw our sacks on the sledge after expressions of thanks and toiled along towards it. As the Eskimos had said, the Lapps were there. Just as we were approaching them, we saw a number of reindeer drawing up behind us, and we soon recognized the sledge party that we had first accompanied out of Unalakleet. The reindeer had become exhausted and the group had stayed all day at a stopping point, so that the reindeer could rest. During that time, we had passed them, but they now had reached us again, and we all camped together for the night. But Brynteson and Hagelin were not with them—they were still back on the trail, and we presumed that their reindeer were too tired to keep up. In the morning, we who were on foot and pulling a heavy load resumed the journey. The reindeer sledges remained behind, for there was good grazing for the deer at the area where we had camped. We needed to rest, as well, but we had to move along as quickly as possible; also it was cold and windy along the open shore, and no firewood was to be found. The wind was with us, though, and we sailed rapidly away, arriving just after midday at Moses Point, where a number of Eskimos live. They invited us in, and offered us room for the night, and in fact we did not want to travel farther that day. We accepted their kind offer, not wanting to sleep under the open skies due to the cold and the wind. I reckoned the temperature to be 20 to 25 degrees below zero. The reindeer sledges arrived an hour after our own arrival, and we invited everyone to have a cup of hot coffee, while the deer had a chance to rest for a few minutes. They left after having coffee, planning to stop about five or six miles farther along the shore, where they could find moss for the reindeer.

## April 16–April 22

— We got up early Sunday morning. The snow conditions were poor, due to frost, but we were fortunate to find an Eskimo with four dogs to help us to Golovnin Bay with our load. We agreed to pay him 13 dollars for this help, and we left immediately. In the evening we again caught up to the reindeer sledges, and we went half a mile beyond them, where we camped for the night at the foot of the mountains; there the trail leads to Golovnin Bay. Again we were up early, and we began the climb up the mountains, which rise rather steeply. By midday, we had gone about one and a half miles, to the crest, whereupon we set the dogs free from the sledge, hopped on the load ourselves, and swept down the mountain slope at a merry pace, all the way to the other side, down by the fjord (Golovnin Bay). There we made some coffee, ate a little food, rested, and labored on our way until we reached Golovin in the evening. We met Jafet Lindeberg immediately, and he showed us into an Eskimo house, where we obtained lodging. Lindeberg was staying at the mission house with Anderson. Our lodging house was owned by an Eskimo woman, a widow, who lived together with her son and daughter. They were nicely dressed, and everything in the house was in good order—clean and pleasant.

During the week we met Carl Suhr, who had left Eaton Station a month before we did and had arrived at Golovnin Bay a few days ago. He had become snowblind during the journey; in addition, he, too, had a heavy load to pull behind him, and he had been alone during the entire time. Since he was snowblind and there was no one to help him, his journey had been long and difficult. The reindeer sledges arrived the day after we did, but Brynteson and Hagelin did not come in until three days later, for their reindeer had become exhausted and it was necessary to let them rest often, which caused their delay.

Golovin town consists of the Mission House, the schoolhouse, the house and general store of Merchant Dexter, and about ten Eskimo houses—all situated on a lovely, low promontory which extends out into the fjord from the east.

## April 23–April 29

— The reindeer sledge party left on Tuesday evening, and Jafet Lindeberg went along with them, while we did not leave until Wednesday morning; we asked Carl Suhr to go with us. We toiled with the sledges all day, until we reached the reindeer party in the evening at Cho-kok; there we stayed in a deserted Eskimo hut in which no one was living for the winter.

The reindeer sledges left at midnight. They had to rely on nighttime snow conditions to travel now, since in the daytime the snow was soft. At night, it was cold and the snow was hard, and also at night there was so much light that one could travel as far as one wanted; one could see almost as well as in daytime. Early in the morning, an Eskimo came to our hut; he knew that we had stopped there the evening before. He made a fire and melted snow for water, so that we could have coffee. He may have been hungry. We invited him to have breakfast with us, and we all ate together. It was still early, and snow conditions were good, and we set out again. But before long, Carl Suhr said that he was having pains in his legs, and that his eyes were troubling him. We slowed our pace so that he managed to keep us with us—we did not want to become separated. We covered a good distance under the favorable condition of the snow. In the evening we even managed to pass by some of the reindeer sledges. We were invited to stay over for the night with an Eskimo family. We observed that the man had two wives, and in fact he told us that the women in the house were his "nuliat" (wives). One of them was young, the other older and not so attractive. We noticed that when they went to bed, the younger of his wives slept beside him and the older was by her side, close to the wall.

In the morning we resumed our journey, although we did not feel quite up to it, having gone at too hard a pace the day before. We went onward until time for dinner, mid-day, and there by the shore we had to stop. We set up our little drill tent on the snow, among some stacks of wood that the Eskimos had collected, and used some for a fire. We stayed there the rest of the day, and all night, with the hope that Carl Suhr's eyes would improve, but they were worse rather than better, in the morning. Johan Petter also did not feel well; he had a cold, which he attributed to the unhealthy heat in the Eskimos' house. He was so racked by nausea that he vomited and coughed up blood. I could do nothing but give him a little medicine and massage his chest with camphor drops, which I had obtained from Dr. Gambell to be used on our trip. We were thus obliged to remain quiet and wait until everyone was well again. But also the weather was bad, and we should not have been able to travel even had everyone been well. We rested all of Saturday. I decided to return, in the evening, to the houses of the Eskimos whom we had passed by yesterday, about three miles from our present camp, with the hope of buying from the people some lingonberries that I could prepare and give to the sick men. As I approached the first hut, where no Eskimos were living, I saw two sledges that looked like those used at Eaton Station, so I peeked down through the smoke hole of the hut, saw that people were inside, and called down to ask who they were. At once when they responded, I recognized Mikkel Nakkila and his wife, Bereth Anna, who were traveling to their claims at Cape Nome. They had left Eaton Station on Wednesday and in three days had come just as far as we had in our travel for almost three weeks. One can realize how rapidly a journey can go when a tamed, experienced transport reindeer is pulling the sledge. Mikkel went out to fetch the reindeer from the area beyond, while Bereth Anna made coffee. Then, after the welcome refreshment, we left together and drove to our camp, where they stayed for a few hours. They departed, planning to reach Cape Nome before morning, a distance of about twenty-five miles. Carl and Johan were better now, and both said that we should try to go onward if we had good weather in the morning, at least until we should reach an Eskimo settlement where possibly we could spend the night; for it was not pleasant to lie in the sleeping bags on the snow, especially when all were not feeling well.

## April 30–May 6

— My comrades felt well enough that we decided to try to move a bit farther ahead along the shore, possibly making our way to the Eskimo houses that we knew lay about eighteen to twenty miles beyond us. We managed to reach them late in the evening, and we put up for the night in a hut where no one was living. Next day, we reached a hut wherein a Norwegian was living; he was the recorder for Bonanza district and had his office there. He said that it was about fifteen miles to Anvil City, where about 300 men were staying. On our way again, after about an hour we came to the mountain which bears the name Cape Nome, which is the boundary marker between Bonanza district and Cape Nome district—Bonanza lying to the east and Cape Nome to the west, encompassing twenty-five square miles. At dinner time we arrived at Nome River, where Regnor Dahl was living. Here, first of all, we met Otto Leinan and Amund Hansen, who after leaving Wilhelm Kjellmann, had worked for Dahl during the winter; they had sawed timber, from which they had built a house for the North American Transportation and Trading Company, on the shore. They had also sawed much wood, which Dahl had transported up Nome River to the tributary streams where he had staked most of the claims.

Carl Suhr stayed here at Nome River, with Otto and Amund, in their little Eskimo hut of sod by the shore. We three others went on to Anvil City, where we arrived in the evening. At once we met some of our friends—Jafet Lindeberg, Magnus Kjeldsberg, Thoralf Kjeldsberg, Wilhelm Basi, and Ole Berg. All of them were well and healthy. We pitched our drill tent beside Kjeldsbergs'. There were tents along the entire shore, for that matter, but most of them were at the mouth of the Snake River on the eastern banks, where people had staked out house lots or town lots for themselves. After Wilhelm Kjellmann and A. Spring had measured, staked, and recorded the area for themselves during the winter, others had claimed ground. Now the place was full of tents, each person with his own tent on his own lot. In addition, they had dragged logs up from the beach and erected them to form houses, which was in accordance with the law of the district; in order to retain his lot, each owner had to bring in at least twelve logs to his site.

We rested from our journey all day on the 2nd of May, just visiting, asking questions, and hearing how everything was arranged here, and which way one should go in order to find himself a claim. We were told that all of the streams were staked in claims out to the very end of the Snake River, but it was presumed that there were streams along the Nome River that were not yet fully taken up in claims. So we prepared to make an exploratory trip up the Nome River the following day. We packed a little food and our sleeping gear on the sledge and went over to Nome River, where Carl Suhr and Amund Hansen joined us, and we set out for the mountains along the river, turning off at Osborn Creek, which we followed, having found the entire river to be staked out in claims. We camped there by the stream, and managed to get some willow twigs to burn, so that we could make coffee. Other wood was not to be found. Bare mountains and valleys rose one after another, as far as the eye could see. In the morning we went farther up the stream, and set up camp in the afternoon, when we observed that no more claims were present in this area, and farther upstream. With spade, pick, and pan, we tried to look for evidence of gold. Finding nothing, we staked no claims. We bedded down for the night, and froze a bit, for it was still very cold. Next morning, we crossed the mountain to Nome River and headed upstream, arriving at Regnor Dahl's tent in the evening. His companion was there, but Dahl himself had gone with a reindeer sledge to look about; he returned late in the evening.

## May 7–May 13

— Sunday morning, we went on ahead, about four miles from Dahl's camp, and we set up our own camp by the river. Again we gathered willow twigs for a fire, planning to continue farther the following day. But Carl and Amund were rather weary of the journey, so they turned back and headed for the seashore. There were only two of us now—Johan Petter Johannesen and I. As we sat in our tent, we heard sleighs nearby, drawn by reindeer, and they came up to our camp. It was Per Larsen from Eaton, with two reindeer, and Alex Joernes with his big dog, Gypsy. Gypsy ran just as fast as the reindeer, with Alex sitting on the sledge. They too were exploring the Nome River and searching for claims for themselves, if there were any places available. So we made plans to travel up along the streams together, continuing the search. In the meantime, Johan and I had found one stream where no one had yet made claims; so we staked out the stream and named it "Darling Creek;" each of us made a claim here. Since Johan and I were not fluent in the English language, we asked Alex to check the bearings on the location and to write up the proper notices. Then we moved our camp up to Clara Creek, investigating the stream all the way to the end; but it was staked out everywhere, so we returned to the Nome River. Each of us made a claim there, just above the mouth of Clara Creek, the first on a small island, and we called this one "Cuba," while all of the other claims were numbered. From Clara Creek we left in the morning, planning to visit Osborn Creek again on the way down. We came to Regnor Dahl's tent in the afternoon, but found that he had driven downstream. Our plan was to reach Osborn Creek during the afternoon, but not until late in the evening did we arrive, for we wandered about for a while in a thick fog. At last we found the way—we recognized our own tracks in the snow where we had gone earlier. We followed the tracks and came to our old tent site. There we made our camp again, and spent the night. In the morning we had planned to do more exploration, but there was such a snowstorm that we could not see ahead, and to travel was impossible, so we decided to stay in this place until the next day, and if possible then to continue the descent. We could have survived for a longer time, but we were a bit short of provisions. Since we had nothing else to do, Johan Petter and I went out and staked claims for ourselves on a little watercourse named by us "Ida Gulch;" I took claim No. 1, and Johan Petter, claim No. 2. On that stream there was no space for more than those two. We did this mostly for amusement, and to keep warm, and also because the sites were quite close to our tent. Next day we decided that since we were about to run out of food, we would head downward in spite of the fog and difficulty in seeing the terrain. We walked from Osborn Creek back to Nome River, but rather than going all the way down to the town, we crossed the mountains to Anvil Creek. In the fog and poor snow conditions (the snow was very soft), pulling the sledge was miserably wretched. At last, late in the evening, we came to Wilhelm Kjellmann's cabin at his claim No. 2 Below on Anvil Creek. There a number of our friends had their tents, also. We were invited to enter and warm up and dry off a little, and to have a bit to eat. We came down to the town in the morning on the 12th of May. There we had our supplies, which we retrieved intact and untouched. The next day, we rested and recovered from the difficult trip in the mountains. Almost everyone was staying in camp; only Jafet Lindeberg's men were at work, hauling up wood and planks that he had purchased from some people. His men had sawed up many logs that they had taken from the shore where the sea had cast them up.

## May 14–May 20

— For nearly the entire week there was foul weather, with rain and wind, so we along with most everyone stayed quietly in camp. Only those who had reindeer to drive could travel sufficiently far to haul necessary materials to their claims, and those were mainly on Anvil Creek; for example, Jafet Lindeberg, who owned the Discovery claim and claim No. 1 Below, and J. Brynteson, who owned claim No. 1 Above, were able to get to those sites. Since I had not much else to do, I gathered some information about the area. On Anvil Creek, claim No. 2 Above was staked for Missionary Anderson; No. 3 Above, for Hagelin; No. 4 Above, for Missionary Karlson; No. 5 Above, for Missionary Huldtberg; and Lindblom owned No. 6 Above. Dr. Kittilsen owned No. 7 Above; Price, No. 8 Above; and No. 9 Above was staked for the Eskimo

Jafet Lindeberg.

*Photograph courtesy of the Alaska State Library Historical Collections, Alaska Purchase Centennial, PCA 20–88,* Discoverers of the Nome Gold Fields.

lad Konstantin, but since the Eskimos could not hold claims, and so that he would not lose it, the claim was held in the name of Price's brother. The rules were thus bent, in order to keep the claim. Johan Sp. Tornensis owned No. 10 Above, Mikkel Nakkila, No. 11 Above; No. 12 Above was owned by J. Hvidsted; Stefan Ivanof of Unalakleet owned No. 12-b; and Ole Olesen Baer owned No. 13 Above, which is the last claim, at the end of Anvil Creek. At the upper end of No. 11, a small stream, named Nakkila Gulch, flows down to Anvil Creek. On the Gulch, claim No. 1 was staked for Isak Nakkila, and the gulch was named for him. No. 2 was owned by an older man by the name of Deering. Sometimes the watercourses were thought of as measured by claims: Nakkila Gulch was only two claims in length. At the upper end of claim No. 6 on Anvil Creek, a small river runs down in a ravine called Specimen Gulch, flowing into Anvil Creek from the east; here claim No. 1 was staked for Wilhelm Basi, and No. 2, for Olai Paulsen. On the west side of Anvil Creek, about at the middle of claim No. 6 there, runs Quartz Creek Gulch, where claim No. 1 was staked for Magnus Kjeldsberg and No. 2 for Thoralf Kjeldsberg. Claim No. 2 Below was Wilhelm Kjellmann's, and claim No. 3 Below was staked for Dr. Gregory in St. Michael. Along Anvil Creek to its mouth, fifteen claims were established below Discovery. Altogether on Anvil Creek, twenty-nine claims existed, of which many were irregular with respect to size. Many were found to be too large according to the law, and fractions of them were removed; such small portions were often only a few feet in length, but nonetheless they were possibly of great value. Some of the above-mentioned lads, who were to leave Eaton Station a few days after we departed, came in on one day this week, all healthy and in good spirits.

My comrade Johan Petter and I agreed to work for Jafet Lindeberg; when we were able to leave the town, we went up to Erik Lindblom's claim No. 6. A temporary camp had been set up, so that effective work could begin as soon as high water receded. First, the ice was to be chopped away, so that the ground could thaw more quickly. At Mountain Creek, Rock Creek, and Dexter Creek, similar work was beginning; the sites at those places had been leased to several parties. We of Lindeberg's group were to work at Anvil Creek and Snow Gulch.

## May 21–May 27

— We went over to Snow Gulch—J. Brynteson, J. D. Morgan, Johan Petter, and I. Morgan was our foreman, in the work to open the claim; here, at this time, only a few were to work, mainly to chop ice at Brynteson's claim No. 1, where about 1775 dollars in gold had been panned last fall. Lindeberg owned No. 2, Lindblom owned No. 3, and No. 4, the last claim, was owned by Dr. Kittilsen. With a number of men, Lindeberg and Lindblom were starting the work at Anvil Creek; in fact, almost everyone had accepted work at Anvil Creek. They were gradually to chop the ice, in order to get under way with the panning, for there was a kind of competition, to see who could be the first to pan out 10,000 dollars in gold from the Cape Nome district. The first one to achieve that amount was promised free passage to the States by the company in St. Michael. But not much could be accomplished this week, as the weather was cold and stormy, sometimes rain and sometimes snow, so that we had to keep to our tents.

## May 28–June 3

— The weather was a little better, and work picked up a bit. Pioneer Mining Company and Price eagerly began to work for that first 10,000 dollars. Price already had taken about half that much from claim No. 8 at Anvil Creek. Brynteson was a few hundred dollars behind at Snow Gulch. Not more than 400 to 500 dollars' worth of gold could be panned per day, although sluicing continued all week (on the days when the sluice box could be used). A lot of claims were jumped, mainly those belonging to the first men who came to Cape Nome, who had the best ones in the district, for example, the Pioneer Mining Company's people, Jafet Lindeberg, John Brynteson, and Erik Lindblom. But many others were also jumped, especially those thought to be good (valuable), thus creating disturbances and uproar throughout the district. If the miners had been able to act according to the opinion and will of the majority, all of the claim jumpers would have been thrown out of the district within a few hours; but orders had been received from the authorities in St. Michael that no legal action was to be taken until after they arrived with some soldiers. After that, the judge would settle all points of conflict in each case where the law had been broken.

Work had begun on Anvil Creek from claim No. 2 Below to claim No. 12 Above; on Snow Gulch on claim No. 1; and on Mountain Creek, where Pioneer Mining Company had leased its claims. Also that company had leased claims on Dexter Creek, and work had begun there, although all of those had been jumped. Elsewhere, work had not yet started. All this week, men went about, going into the mountains with spade and pan on their shoulders, since most of the snow had been melted away by the rain; only the ice remains in the streams, and the sun will soon take care of it. Down in the town, people were standing about, waiting until the ice loosened and receded from the beach, so that a channel would open for ships to come in, especially with supplies and materials of different sorts. Many were down with scurvy, and a number had been taken to Golovin and St. Michael, where they could get medicine and better care. It seemed that many with scurvy were Frenchmen.

## June 4–June 10

— The ice went out from the shore on Sunday, and a good breeze blew at sea. The next day a schooner arrived from Unalakleet. On board were passengers who had heard of the gold found at Cape Nome, and they had sailed on the first ship, to arrive as soon as possible.

The claim jumpers have been very busy; some claims had been taken illegally by several people in succession. These scoundrels found themselves in another trouble; they presumably went into a partnership, about twenty of them, as far as I can ascertain, led by men from Chicago who caused all of the uproar. They had gone over to the Sinrock River, west of both the Penny River and the Snake River, an area belonging to Cape Nome district. There, where numerous claims had been staked out on streams that had relatively good prospects for gold yield, these men had set about forming a new district. The man they selected as recorder ordered that all who had claims in the area should report to him, to record their claims with him, and if they did not do as he had ordered, the claims would be taken (which meant that the claims would be jumped) by his group—but all those claims had been recorded already at Anvil City. As soon as they became aware of the actions at Sinrock River, the miners at Anvil City held a meeting, and the majority voted against the men at Sinrock River, since that river was within the Cape Nome district. The Anvil City miners gave the Sinrock recorder only ten hours' consideration: if he did not abandon his post and leave within that time, he would be forced to do so. The man complied, and the scheme was completely stopped; nothing came of the new district that was to be formed.

This week, most of the men were at claim No. 1 at Snow Gulch and No. 6 and No. 8 at Anvil Creek, prepared to begin work, but waiting until the thaw was more advanced; the streams were still completely ice covered, but water was running over the ice. The weather was relatively good.

## June 11–June 17

— It was lively all week. Since there was good weather, a lot of work was done. Price was the first to pan 10,000 dollars in gold at his claim. But it was said that he had borrowed some "dust," since he had been short of the total sum by a few hundred dollars, and Pioneer Mining Company was about to pass him in this competition. The company probably would have won the trip to St. Michael (and to the States) if their foreman had realized that he should have opened work also on a placer claim. Up along the streams, there was much activity, and it was also lively in town, where a number of sailing ships had arrived, loaded with supplies and alcoholic spirits. Saloons, barber shops, and restaurants were set up in tents, and these were the beginnings of life for Anvil City. From St. Michael came the North American Transportation and Trading Company, and the Alaska Commercial Company, which purchased building sites or lots from Wilhelm Kjellmann. On their locations, they set up their stores in tents and filled them up with goods for sale to the miners. At first, flour was at a price of $20 per sack (25 kg), roast beef was 50 cents per tin (1 kg), and thus were seen the prices on all goods, according to the increased rate. A schnapps at the saloons cost 50 cents, as did a cigar. A shave was 50 cents, a haircut was $1; a day's wages up at the mines was $5 to $7.

## June 18–June 24

— Cold weather came from the north. The ice melted more slowly and the earth was frozen everywhere. We had no sunshine all week, but rather snow and rain. More materials had come from St. Michael and the States, so that the work of building houses, mainly for stores and saloons, was well underway. We heard a rumor that about thirty women had arrived at Anvil City, from Dawson City, to begin their profiteering in the town; and it was said that almost at once upon coming ashore some of them went inland to look for claims. By now, scarcely any unclaimed ground is to be found, especially on the streams closest to Anvil Glacier, but also on other nearby watercourses; they were all staked out from mouth to source, and many were double-staked (that is, jumped), so that one had to walk about thirty or forty miles to find areas still available, and the women probably were not strong enough for that. Since they did not know the situation, they thought that there was space left everywhere, which was not the case.

This week, Magnus Kjeldsberg began work at his own site, together with his brother, at Quartz Gulch. Since there was a water shortage, he was able to lease claim No. 3 at Snow Gulch for 33 per cent of the returns.

## June 25–July 1

— A number of soldiers, including a lieutenant, arrived, sent by the U.S. Army captain at St. Michael, to keep order in town and to settle all points of conflict until the judge himself should arrive. A lot of the workers at Snow Gulch collected their pay and left, so John Brynteson was often short of help, perhaps his own fault for pressing too much out of his men. His desire for riches was really at fault, and not he himself, for he is a man who does not transgress the word of God. Often it seems that when a person acquires much money after having had nothing before, the joy of the money may draw that person into feelings of independence and loftiness. Jafet Lindeberg began working with a number of men at Discovery Claim this week, and it is to be opened for the summer season. Many people arrived from the States on the steamships *Roanoke* and *Garonne*. Some of the crew deserted their ship and came up to Discovery Claim, where they took work with Lindeberg, and some were employed at Snow Gulch as well. An agreement on wages set the rate at $8 per day, and meals to be included. If anyone wanted to prepare his own meals, he would receive $10 per day. One who wished to pay only for meals was charged $5 per day.

## July 2–July 8

— This week the weather was relatively good. Numerous recent arrivals went about and looked around in the mountains nearby, across the river, and at the terrain in general; as they passed by Discovery we noticed them, men and women both, some on the way up and others descending. Work was lively all along Anvil Creek, from claim No. 2 Below to No. 12 Above, and everyone had set up sluice boxes, ready to receive sand from the creek. In town, construction of houses is going on without delay, and many of various sizes have been built already. Most prominent along the streets are the saloons, none yet ready but to be completed before the end of next week. I noticed that some of the women in town dressed as do men, and they do look rather odd. One of them was working along with some men, floating materials ashore and up the Snake River to the mouth of Anvil Creek and even up to Glacier Creek. These people made quite a lot of money that way; all such work was done under contract. In the town, several large companies have arrived, using tents for their stores until their buildings were ready. Aside from Alaska Commercial Company and the North American Transportation and Trading Company, the largest establishments were Alaska Exploration Company, Sampson Company, Cleveland Brothers Company, and Cape Nome Mining Company, but other, smaller enterprises were present now, too, so that stores and goods for sale were everywhere along the beach.

Tuesday was the Fourth of July. In Anvil City on that day, there was scarcely any one around, for people were as if half wild to get gold claims for themselves or to set up business, whichever one they were planning, and almost everyone was away. Not many people were remembering to honor the Fourth of July. I did see a few staggering about rather drunk, but all else that I observed was work traffic. I did not hear very much talk of the meaning of the Fourth of July anywhere.

## July 9–July 15

— A meeting of the miners was held at the beginning of this week, called in an attempt to get the length of claims shortened from 1320 feet to half that amount, as in the Klondike in Canada's territory, and as provided in Canadian law. The mining laws of the United States, on the other hand, state that a person could stake 20 acres of the government's land. But when the parties began to dispute and disagree over the proposal to change the size, the lieutenant had his men chase everyone out, and thus the meeting was ended. We knew that the men who wanted to have the claims at Cape Nome reduced had all come in from Dawson. The brawls and uproar began over this matter as soon as those men from Dawson arrived. Later in the week, another meeting on the same subject was convened, but it also was ended by the soldiers, with no results, and no further meetings were to be called with regard to reduction of the claims, so the Yukon men did not get to cut in that way but they had to try something else.

Some men have taken a fraction of claim No. 1 at Nakkila Gulch, at the lower end of the claim. They had no right to take the portion at the lower end, since the claim was staked up the gulch. In actuality, the claim was too long, but the fraction of overage should have been taken at the upper end. These people evidently thought that they could do as they wanted with the Lapps who own this claim, and they actually went over and jumped it entirely. Isak Nakkila has been in court all week, and he recovered his claim, but he lost the upper fraction. Claim No. 9 at Anvil Creek also was jumped, and the men involved appeared on the site and began to pan and to use a rocker, but they were all taken down to the town by the soldiers; the soldiers had come up to one man's tent just as he was fixing supper. He did not have time to finish, but had to leave everything and depart the claim at once. The soldiers then broke the rocker into pieces and annihilated the camp.

## July 16–July 22

— On Sunday a strong storm blew in against the beach, with heavy seas that broke and sprayed up over the sandy strand about 300 to 400 feet from shore. Two large schooners that could not get under sail in time were driven ashore. A small steamboat, a pleasure boat, and a large praam were cast up on the sand and were left dry on the beach, but no human lives were lost. All of the steamships and sailing ships that did manage to get underway went to the lee of a small island about twenty-five miles farther north, which lies only a few miles offshore, thus forming a sound between the island and the mainland. There the ships have good harbor from wind and storm.

The Pioneer Mining Company's people carried down to town this week the first gold they have taken out—worth $37,000. Nothing has yet been panned from No. 6 and Discovery claims at Anvil Creek. No. 6 does not appear to be so rich as we had thought, and Discovery and No. 1 Below both require much labor to open for the season. Work is underway on all the claims at Anvil Creek, but sluicing is not possible everywhere, since the ground is still quite frozen and must be given time to thaw. The weather is constantly mixed with wind and rain, which make it wretched to stand and work during the long days. Brynteson was dissatisfied with Morgan, his foreman up to now at Snow Gulch, and he took a Norwegian, August Lie, for the job; Lie has shown himself to be the man for this job. Also a change of foremen occurred down at Discovery, where Lindeberg replaced old Mr. Deering with Dr. King. Deering went as foreman to No. 6 claim, after Lindblom dismissed Julius Fold, his foreman there.

## July 23–July 29

— Dr. Sheldon Jackson arrived with the steamship *Bear* from his trip to Siberia, where he had gone this spring some weeks ago. The *Bear* had picked up Pastor Doty and his companion, Jeremias Abrahamsen, at St. Lawrence Island, where they had spent the past winter, after having been sent there in summer 1898 by the American Missionary Society to teach civilization to the Eskimos of the island. On the *Bear* also was an Eskimo who was to be taken to St. Michael or to Sitka for punishment of some sort, perhaps to

be imprisoned, for the attempted murder of Doty and for many other abuses that he had devised. We were told that he would never be permitted to return to the island, where he left a wife and children. Pastor Doty was to travel to his home in the city of Washington; Jeremias Abrahamsen came ashore at Anvil City and started working for Jafet Lindeberg immediately.

Dr. Jackson sent for Wilhelm Kjellmann at Anvil Creek, and together they left for Unalakleet. Jeremias told us that he had met Dr. Jackson at once upon boarding the steamship *Bear*, and they talked together about the Reindeer Expedition and about Unalakleet, but Dr. Jackson did not say anything about Cape Nome nor mention that all of us had left his service and had gone to the gold fields. Dr. Jackson must have known about the situation (at Eaton Station and Cape Nome), for before Christmas time, Wilhelm Kjellmann had sent him by the overland mail a report with full information, which Dr. Jackson should have received long before

Sheldon Jackson and others of the Department of the Interior, at Teller Reindeer Station in 1901. From left to right, the adults are: Francis H. Gambell (physician), T. L. Brevig (teacher/missionary), A. Hovick, Mrs. Brevig, Sheldon Jackson, Mrs. Campbell, E. O. Campbell (physician), and A. R. Cheever.

*Photograph published in Sheldon Jackson's report, 1903.*

he left the States. Abrahamsen was at a loss to know why he had not been told about all of this, and that Dr. Jackson would not need him in his service any longer. He did not know at all about the situation that had arisen last winter regarding gold at Cape Nome.

At his claim No. 1 Below, Jafet Lindeberg found that the first collection of gold, after three days of accumulation of sand by six of his men, amounted to a value of $5000.

Along the shores of the Snake River many river steamers have anchored, and the people aboard are conducting business—some have grocery stores and others have hotels and saloons. The steamship *Sadie* arrived from St. Michael, and I could see the people, men and women, coming ashore. Some of them were rather old, white-haired, walking with canes, often appearing bent and a bit shrunken, and not able to walk well; but it seems that money is still on their minds, although they seem to have one foot in the grave. Alaska is anything but a place for old people to live, for life here is difficult even for the young; young men have streamed in, happy and healthy, over a number of years, having gold fever, leaving their good homes elsewhere to go to Alaska to gain wealth. Most of them I judge have found it not an easy matter to gain wealth here, for many, after they have been here for a while and used up all the money they had brought along from home, must wish to be back home again. I have heard some people cursing themselves and Alaska, that they should have been so foolhardy as to travel here and find only grief and want. Some have also gone to their graves here, although they set out so vigorously and healthy.

The weather was excellent all week. The warmest day was 86 degrees. News came that people were panning gold down by the seashore, and the rumors were so convincing that men left their work along the streams and swarmed down to the beach, where they constructed rockers for themselves; indeed, they took out great amounts, from five to twelve dollars in gold per man, and even up to thirty dollars. As far as I know, the discovery of gold at the shore can be attributed to some Frenchmen, who were maintaining last winter that one could pan out per day five dollars' worth of gold by the shore of the sea. They had only been talking and guessing, but their suggestions have turned out to be so.

## July 30–August 5

— Everything was peaceful and quiet this week with respect to disturbances, for almost everyone was busy panning for gold at the beach. There is a shortage of laborers at the mines on the streams. In town, there is activity; ships have come and gone every day, with people and supplies. Supplies do not meet demand, and prices are high for materials. Poor people are not able to buy things and are spending the summer in tents, with plans to go down to the States in the fall, for the winter. Lumber at the cheapest is thirteen cents per foot, $130 for a thousand feet, and up to $250 for two thousand feet. Perhaps the prices will go even higher, due to the demand. All week we had wet weather.

## August 6–August 12

— At claim No. 1 Below (Anvil Creek), a lump of gold worth $312 was found. The millionaire, Mr. Lane, has sent ten horses to Cape Nome; they are hauling supplies up to the mining camps every day, primarily to Anvil Creek and Dexter Creek and Snow Gulch, where the largest operations are. Last year, Lane had agreed to furnish supplies to, or to outfit, Price in the operations here, with the condition that he would have half of whatever good fortune Price would find in Alaska. Now, Lane owned half of claim No. 8 at Anvil Creek, as well as at Dry Creek and other streams where Price had staked claims during the winter. The work at Mountain Creek has ceased, since the men there have not found the prospecting to be rewarding. People were at work at Rock Creek, where prospecting was relatively good, and at Glacier Creek. From Nome River area, we heard very favorable reports about results at Preston Creek and Osborn Creek. None of us had heard from Regnor Dahl, who had staked claims for the North American Transportation and Trading Company farther up by Nome River. At Clara Creek, good prospecting had been anticipated, and that turned out to be the case. Saunders Creek is a branch of Hastings Creek, which runs into the sea about four miles east of the Nome River. I tried my hand at prospecting there and in one pan found twelve large pieces of "color," like large lumps of sand. But since my companion and I did not have supplies enough to stay there, we had to go back to our regular work at Anvil Creek; we met on our way a lot of people who were going to Hastings Creek to make claims, but actually, that creek had been staked out and taken since early in the spring.

## August 13–August 19

— Yet another dispute is going on, started by some deceitful people. They questioned whether Lapps had any more right than Eskimos to hold claims, and whether the judge from St. Michael could legally issue citizenship papers. Many from the expedition had taken out their citizenship papers in St. Michael, including Jafet Lindeberg. There was also a question of whether people could hold claims through powers of attorney. These questions were to be decided when Governor Brady and a judge arrived from Sitka, from where they had been called. The disputes at Cape Nome district could not be settled otherwise, to everyone's satisfaction, it seems, until that review takes place.

## August 20–August 26

— The governor and the judge arrived the first of this week. Governor Brady and Captain Walker (U.S. Army) went up to Lindeberg's camp at Discovery, where they spent the night, returning to town next morning. All week, the court was in session, ending in some good results: The Lapps retained their claims and the powers of attorney were found to be legal; one could stake claims for a companion, even if the power was given orally, and in the presence of witnesses, should a written power of attorney not be obtainable. The court found, in a sense, against the question of citizenship papers' being issued by the judge at St. Michael—that he could not legally issue the citizenship papers that he had been stamping and validating. But those who had received their citizenship papers in St. Michael were permitted to keep their claims as genuine citizens, for they had applied in good faith and could not do better at the time, and the court found none to be at fault. In disputes over claims, only one claim jumper was permitted to retain a contested site, when his opponent in court could not prove that he had staked and recorded that claim prior to the time of the secondary claim (that of the claim jumper).

There was rain and wind all week.

## August 27–September 2

— Since the nights are getting dark, robbers and thieves are finding it easier to go about their business. In town, an attempt was made to rob a lad who lived alone in a tent, as he was assumed to have some money. When the lad denied that he had any money, the thieves gave him several blows on the head with a revolver, but they left when they found that he had nothing. On Anvil Creek, gold reckoned to be worth about $400 was taken from the sluice box at claim No. 11 during the night, while the miners were inside the tent, eating. In somewhat the same manner, about $800 in gold was stolen from the sluice box at claim No. 4, which was Missionary Karlson's. Snow Gulch was also robbed—by the watchman at the site. One of the workers saw the watchman taking the gold and reported this to Brynteson, whereupon the watchman was forced to turn it over at once; it was valued at fifteen hundred dollars. With all this going on, a meeting in town was held, with the purpose to obtain an agreement that all thieves and robbers were to be thrown out of the district, as far as they could be recognized and proven to be such. But the meeting ended without instituting that plan, for it was asserted that many innocent people might also be sent away from the district. It seems that this was to the advantage of the thieves, who presumably would want things to stay as they have been, with no place safe.

## September 3–September 9

— While many of the camps were thus being visited by robbers, Discovery's turn came up as well. On the 5th of September, at 2 o'clock in the morning, the watchman saw a man come out of the office tent. The watchman approached him and asked, "Is that you, Lindblom?" The man replied "Yes," and the watchman thought it was, as he knew that Lindblom usually slept in that tent, as did Dr. F. King and Rolf Wiig. The man turned into the tent, saying "Good night" to the watchman, and he went inside. The watchman also left, thinking that Lindblom had been out for a night's stroll and now was returning to bed. After about three minutes, Dr. King shouted for the watchman, saying that the camp had been robbed. The money chest had been taken. The watchman immediately ran a distance up from the tent and came to three men by the ditch. He detained them while he called for Dr. King, who answered that he would be there in an instant. At this, the three men decided that it would be foolish to wait, and they leaped across the ditch. One of them fired two shots at the watchman, but didn't hit him.The watchman then fired at them—six shots—but he didn't know whether he hit any of them or not. Some of the workmen, however, heard a voice from the dense shrubbery: "O, Lord Jesus!" But nothing further was heard, or found later. After those men had run off, Dr. King returned to the camp and commanded, "Everyone up, the camp has been robbed, bring your guns along!" At this, we were all on our feet with revolvers and rifles, firing away in the darkness into the thickets, where Dr. King himself was bombarding with a rifle. He and Lindblom then took a torch and several men together went among the hillocks; the chest was found, lying about thirty feet from the office tent (from where it had been taken). The chest lay undamaged. At its side was a pick, which evidently had been for the purpose of breaking the chest apart. The chest was carried back into the tent—and nothing was missing from it; the gold within it was worth about $17,000. We reviewed what had happened and how Dr. King had been awakened. The man whom the watchman had spoken with was the robber, and not Lindblom. He had entered the office tent to hand the chest out to his companions, through the tent wall, which we found had been slashed right where the chest was standing, and then he had gone out the door of the tent. He must have been happy that the chest had been removed so easily, and probably he was hurrying, but he happened to trip on one of the tent's corner stakes. That caused the tent to move, and Rolf Wiig awoke, and saw at once that the chest was gone. He immediately awoke Dr. King and told him that the chest had been stolen. At that, Dr. King and Lindblom jumped up. Had the chest been lighter, the thieves may have been able to get completely away with it, but it weighed about 200 pounds—too heavy to carry on a run. Dr. King found that the blanket where he slept had been rubbed with water. There seemed to be no sense to that, unless the plan perhaps had been to heat up the mattress with some hot water, so that in his bed, Dr. King would feel the warmth on his back and would more easily fall soundly to sleep? Suspicion for this fell upon the night cook, and he was fired and had to leave the camp. No further investigation was made, although it was clear to us that the people who attempted the robbery were very well acquainted with all conditions in the camp, since they had been able to carry out their plans so unnoticed by us.

## September 10–September 16

— Almost the entire week we had clear, cold weather. Ole Rapp, Lauritz Stefansen, and A. Hansen returned from a prospecting tour on Seattle Creek, which lies east of Osborn Creek and flows into a lake near Cape Nome Mountain. Since they found good prospecting by panning, with from five to twelve pieces of gold in each pan (but not in the bedrock), the creek should certainly be of some value. Hansen and Stefansen had located the stream and had the first two claims on it, as well as Discovery and claim No 1 Below. They named the creek after the city of Seattle, where Hansen has his home. The entire watercourse was staked out at once by us lads from Norway; altogether, there were thirteen claims.

On the 14th of September it snowed for the first time since last spring, and we are reminded that winter is near again. On the 13th, frost was on the ground and ice was on the water, but the sun melted all away during the day.

## September 17–September 23

— The men who had been working farther up in the mountains came trudging down with their sleeping gear on their backs and went down to the town in great flocks like migratory birds, for winter is approaching. Now, not much work can be done, for with this cold, the water has frozen. Work has ceased at Snow Gulch, Dexter Creek, and all of the other places, including Anvil Creek, except for Discovery Claim and claim No. 1 Below. Jafet Lindeberg has got sick, with a fever, as have also Wilhelm Basi and Olai Paulsen. It is called typhoid fever, and many people in the town have this same sickness.

The Lapps have sold their claims No. 10 and No. 11 at Anvil Creek, No. 1 at Nakkila Gulch, and two claims at Dry Creek for the amount of $10,000 (dirt cheap). It may be that the Lapps have their best friends to thank for these dealings, for their properties were sold for only a quarter of their real value, to Mr. Lane, the millionaire. He also bought Regnor Dahl's and Ole Krogh's claims on the Nome River. Lane already owned a large part of Anvil Creek, including claim No. 2 Below (purchased from Wilhelm Kjellmann for $15,000), and claims No. 8, No. 10, No. 11, and No. 1 at Nakkila Gulch.

## September 24–September 30

— Jafet Lindeberg, Wilhelm Basi, and Olai Paulsen were taken down from Anvil Creek and were placed in the military hospital this week. Also this week, a lot of people, about 2000 men, who had come down the Yukon River from Dawson, arrived from St. Michael. Now nearly every evening someone is robbed of money, and is beaten and mistreated. A woman who lived with her husband in town was found to have been murdered in their tent, while her husband was at the saloons. With so many new arrivals, work is in full swing to construct houses for those who are planning to stay over the winter. Already it is very cold—too cold to live in the tents—and the snow extends down to the sea. In the mountains, the snow is to a depth of about two feet, and so work has stopped everywhere except at Discovery and claim No. 1. Nearly all tickets have been sold for passage on the steamships going south to Seattle and San Francisco. The steamships *Roanoke*, the *Laurada*, and the *Cleveland* will carry passengers, plus many smaller steamships that are headed for Seattle. Those bound for San Francisco are the *Portland* and the *Bertha*.

## October 1–October 8

— Now this week all work has ceased, at the mining camps along the streams and by the shore of the sea, so everyone has come into the town, some to wait until the ships are ready to leave. This week is therefore a great time for the restaurants and the saloons. People went from one shop to another, and in the saloons drank and gambled every night; the saloons are full of all sorts of gaming boards. One morning, I watched as a man played and lost over $300 in less than five minutes. When that money was gone, I thought that it was plain to see that it was all the money that he had, and he turned away from the gaming board and went off, while the gambling master happily swept the money into a drawer. Buying meals at the restaurants cost about six or seven dollars per day.

The government steamship *Bear* has orders to carry south the people who have nothing with which to purchase tickets, and 250 poor and some who are ill have signed up to be transported on that ship to the States. But unfortunately the ship cannot take more than 150 of these people at Anvil City who have signed for passage. The fever attacked many more this week; about a third of the population here is sick. Every day, one or two have died. The weather is bad at end of this week.

## October 8–October 14

— Five deaths occurred on the 8th, Sunday. On the 9th of October, the new newspaper in Anvil City came out, called the *Anvil News*. In it one can read all about conditions of life in the town and in the district—the paper tells of the events in the saloons and the dance halls, and it gives other news from among the townspeople, and also about the sickness. As well, there is news about gold—how much has been taken out of the whole district, to the total of which Anvil Creek has given one and a half million dollars, and the streams flowing into the Snake River and the Nome River have given three-quarters of a million; altogether, the total is three and a half million dollars. But that figure is probably too high, and two and three-quarters million would be more like it. Of that amount, the Pioneer Mining Company had panned out about six hundred thousand dollars in gold from the various claims worked during the past summer.

We also learned that Mr. Lane, who is from San Francisco, had purchased claims on various rivers, including the Snake River and the Nome River, for about two hundred thousand dollars. At the beginning of this week, he went down to San Francisco on his own ship—the same one on which he traveled north to Cape Nome. One of the owners of the Cleveland Brothers' Company (one of the brothers) shot himself with a revolver one night; no cause could be found for his suicide.

## October 15–17

— These days begin the 42nd week of 1899, during which I have kept this journal.

On its way north, the steamship *Laurada* sprang a leak on the Bering Sea during a hurricane-like storm. The ship reached an island, and all the people were rescued, but it sank with the entire cargo of goods, primarily supplies of food, destined for Cape Nome. The steamship *Portland* arrived with some of the people who had been rescued from the *Laurada*. This week, the fever has not struck so hard as in the previous week, for the doctor has ordered everyone to boil their water before they drink it. Many who had purchased tickets to take passage to the States remain in bed with the sickness and must give up the trip until later, when the infirmity eases and they feel better.

## October 18

— The journey down to the States:

Jafet Lindeberg was put aboard the steamship *Roanoke*, which is to depart on the 19th of October, and Dr. Miller, the military physician who has been caring for him in the hospital, will go with him. (He left Dr. Southward in his place at the hospital.) Wilhelm Basi had recovered enough that Dr. Miller has said that he could travel, but Olai Paulsen is still not well and has to stay behind. Others of my friends also must stay here—among them, Thoralf Kjeldsberg, Magnus Kjeldsberg, and Lauritz Stefansen, all of whom are very ill. Mikkel Nakkila also stays behind, for Bereth Anna lies gravely ill with the fever.

With others, I went aboard a praam, which Alaska Commercial Company's workmen towed out into the harbor, stopping alongside the *Portland*, where we boarded, about 400 passengers. We departed as soon as everyone was on board, and we were looking forward eagerly to experiencing a warmer climate. About twenty-five passengers on the *Portland* were ill, with varying degrees of fever. Many of the sick were also taken aboard the *Roanoke*, which was bound for Seattle, and aboard the steamship *Bertha* were to go a number who had been stricken with fever, to depart Cape Nome for San Francisco three days after we had left on the *Portland*.

Carl Johan (left) and Wilhelm Basi, circa 1899.

*Photograph by Snodgrass studio, Astoria, Oregon.*

On the eve of the 22nd of October, we arrived at Dutch Harbor; there one of the patients died and was put ashore, for burial there. We had about twenty-four hours while the ship was at berth, taking on water, coal, and other supplies. Then we departed, and as soon as we came through Akutan Pass, we had a favorable wind from the southwest. With that, the *Portland* set all of her sails—six in number, four foresails and two square sails. She made then up to eleven or twelve miles in speed; in calm weather, her speed was seven or eight miles.

The wind stayed with us, and we had calm weather and fog the entire time until the 1st of November. Early in the morning on the 2nd of November we anchored in San Francisco harbor, where the inspectors came aboard. We were not permitted to land at the wharf, but had to go across the sound, where our ship lay in quarantine for half a day, until further inspections could be made. No contagious diseases were found on board, and the *Portland* put in at the wharf. All of us who were well enough to stand up on our feet got to go ashore. Each one then went to his own destination in the city, according to one's choice. We met Jafet Lindeberg, who came from Seattle by railway, and he was quite well now. Dr. Miller, Dr. Kittilsen, and Wilhelm Kjellmann arrived at the same time.

We stayed in San Francisco for three weeks, seeing all of the sights. In the meantime, Thoralf Kjeldsberg had recovered his health, and he came south from Cape Nome to Victoria, in Canada, by ship, and from there by railway to San Francisco. He planned to start out within a few days on a journey to Norway, with his brother Magnus, and with Jafet, to visit their homes. Johan Tornensis was headed for Iowa, where he would visit his uncle, who was a pastor there. The rest of our friends were going to stay in San Francisco, while Wilhelm and I were going to travel to Ilwaco, a small town in Washington state. We planned to stay over the winter, until the beginning of spring when we would return to San Francisco and go from there by steamship back to Alaska.

# ANNOTATIONS
# The Journal of Carl Johan

The following notes are arranged by dates, according to the sequence of events described in the journal. References cited are shown in parentheses by source and date; complete citations are given in the bibliography.

On 19 July 1898, the adjutant general requested Lieutenant Devore to inform him of the number of people from Finnmark who were passengers on the *Manitoban*, entered the United States, and were present in Seattle after the trip on the railroad. Devore gave the official count: sixty-eight men, twenty-six women, and nineteen children had gone aboard the ship at Bossekop; since some of the people had left the trains in Chicago (by prior arrangement), he did not know the exact number present at the end of the train trip.

## 1898

**4 February.** Wilhelm Basi also kept a journal of interest. (Finnish Amer. Soc., 1971.)

Two coastal pilots, a customs officer, and Lieutenant D. B. Devore, U.S. Army, went ashore at Loppen (Loppa).

**11 February.** When the author used expressions such as "entertaining" and "lovely" applied to the third deck, he must have intended irony.

**14 February.** Jackson wrote that he looked into the steerage area, found it wet and filthy, and felt sorry for the passengers. (Jackson, Travel Journals. 1897–1898. Presbyterian Hist. Soc., Philadelphia.) Neither Jackson nor Kjellmann could have done anything to improve the situation on the ship, but had Jackson expressed orally to the passengers his regret for their rather miserable conditions, the people undoubtedly would have been happy for such support. When Lieutenant Devore chartered the *Manitoban*, obviously he did not regard the Finnmark people as needing more than only barely adequate accommodations shipboard.

**15 February.** Captain Braes and First Mate Buchanan may not have been on the bridge, but that they were negligent is unlikely. One source of fresh water for the deer was snow; in their concern for the animals' welfare, the men from Finnmark collected snow from the deck after the snowstorms at sea, and the deer eagerly ate it, evidently before it melted. (Jackson, 1898b.)

**20 February.** At least the bread was good. John Hunter, from Alva, Scotland, was the baker. (Board of Trade, 1898.)

**24 February.** A written agreement between the War Department and Allan Line was not found in the National Archives. As on all vessels at sea, cleaning the ship would be a responsibility of the crew, but the War Department actually paid Allan Line an extra amount to care for the deer on board the *Manitoban*. (Alger, 1899.) However, that Jackson received money so that work by the people from Norway would exempt the crew from its duties (as noted, ordinary but on this voyage specially paid) is improbable. Carl Johan later added that he could well believe such a payment after observation of Jackson's unreliable treatment of the members of the expedition and after learning "how he dealt with the poor and unenlightened Eskimos." He noted that Jackson was the "appointed head of education (instruction in civilization) for all of Alaska." Evidently, to Carl Johan (and his comrades, some of them, at least, whose opinions he heard and obviously honored), a promise, written or oral, was a formal agreement; no one was to be excused for failures to carry out pledged intent.

**26 February.** The ship's surgeon was Forbes Hall Wolfe of Glasgow. (Board of Trade, 1898.)

**27 February.** The *Manitoban* anchored in the bay just to the west of Sandy Hook, New Jersey. (Jackson, 1898b.) There the Camp Low Federal Quarantine Station was located. The first inspectors to arrive probably were officers of the U.S. Marine Hospital Service (renamed in 1902 the U.S. Public Health Service), who were responsible for health inspection of all immigrants (mandatory under the National Quarantine Act of 1893). (Furman, 1973.)

Until the vessel was granted pratique (28 February), probably only federal inspectors were permitted on board. Late in the evening of 27 February, the U.S. Army steamship *General Meigs* came longside, with Lieutenant Devore, Lt. Col. A. S. Kimball (Depot Quartermaster, New York), Lt. Col. E. W. Weston (Chief Commissary Officer), Alfred W. Gumaer, J. J. Rafferty, and Capt. Luther S. Kelly. All had official duties with the relief mission. Devore had been ordered to accompany the transcontinental trains as chief responsible officer, since Dr. Jackson was to report immediately to the Department of War in Washington, D.C. The Departments of the Commissary General and the Quartermaster General arranged all details concerning the journey by train.

Colonel Weston had seen to the purchase of necessary provisions for the personnel, for use on the trains; food for eighty-seven"Lapps" was specified in his orders, but the rations probably were increased to meet the needs of the greater number of people actually present.

Transportation of the reindeer was the responsibility of the quartermaster general (QG); commanding that echelon in the District of New York was Colonel Kimball, who was to over see arrangements for railway travel for deer and personnel from New York City (actually Jersey City) to the west coast (Seattle, Washington), as ordered by army headquarters in Washington, D.C., which by the first week of February had determined the plan of action. Beginning in mid-January, the QG invited railroad companies to offer their services by submission of bids, and the Pennsylvania Railroad Company won the contract. The company received $11,204.20 for carrying the deer and personnel from New York to Seattle. (Alger, 1899, page 45.)

The U.S. Army was well familiar with rail shipment of its livestock, i.e., horses, mules, burros, and cattle, but the QG was at a loss to know how to handle reindeer; Acting Secretary of War Meiklejohn had to ask Dr. Jackson by cablegram to Alta how many reindeer would fit in a standard eight-by-thirty-six-foot railway car. Jackson cabled back (and wrote a letter additionally) that a reindeer such as was purchased in Finnmark was about the size of a burro and that the freight agents of railroads in the southwest U.S. would be able to tell the department how many burros could be placed in one car. He also reminded the department that the horns (antlers) would not be present on any of the deer. Nonetheless, with unexpected enthusiasm and interest, the Pennsylvania Line had modified the design of the standard livestock car, to accommodate better the reindeer, and had sent on 1 February 1898 a description and a blueprint, and even a sample of the new reindeer car (the Circle F car) to Washington, D.C., for the acting secretary of war to inspect. For the long journey across North America, the Pennsylvania Railroad made subcontracts; Pennsylvania's lines would carry the deer and accompanying personnel from New York to Chicago, there to transfer to the rail lines of the Chicago, Milwaukee and St. Paul Railroad, and at St. Paul to the rails of Great Northern into Seattle. (War Department, AGO File 67221, National Archives.) The two trains together were reported to comprise twenty-two cars holding the reindeer, one carrying sledges and other equipment, eleven with reindeer moss, two for personnel, and two dining cars. (The *Chicago Tribune*, 3 Mar 1898, p. 4, and the *New York Times*, 28 Feb, p. 5, each gave slightly different numbers.) Evidently, Dr. Jackson kept no personal notes about the relief mission after his return to America, and Lieutenant Devore's messages to the War Department reported only dates of departure in New York and arrival in Seattle.

The War Department had three of its own professional guides or scouts aboard the trains. A. W. Gumaer, from St. Paul, Nebraska, was appointed special agent of the War Department, to superintend care of the deer on the journey to Seattle. His assistant was J. J. Rafferty, from New York City, also hired as special agent. Capt. L. S. Kelly (known as "Yellowstone Kelly" from the days of his service as chief of scouts for General Nelson Miles in the campaigns against the Dakota people) was assigned to the expedition under special orders. (Greene, 1991.) Gumaer and Rafferty were paid salaries and were authorized to draw military rations en route. Dr. Arthur Trickett, to board the trains at St. Paul, Minnesota, was appointed veterinarian, with no salary but permitted to draw rations. (War Dept., AGO File 67221, Nat. Arch.)

**28 February.** Regarding other mandatory inspections of incoming vessels and their passengers, commodities, and freight, Dr. Jackson suggested to Secretary Alger that he ask the secretary of the treasury to issue instructions to the collector of customs at the Port of New York to pass without detention the reindeer, the sledges, any herding dogs present, and the "Lapps" with their personal property; also that the secretary of agriculture curtail the customary quarantine of livestock from Europe (in this case, the reindeer). The collector at the Port of New York was requested to notify the *Manitoban* upon arrival as to the pier assigned her, so that personnel and deer could be transferred to railway cars without lengthy barge travel. The Department of War was assured of cooperation from all agencies. (Bur. of Animal Husbandry, Dept. of Agriculture, File 3845, Nat. Arch.; War Dept., AGO File 67221, Nat. Arch.)

The vessel *Thingvalla* (Thingvalla Line, Denmark) departed from Kristiansand, Norway (Captain Laub in command), on 10 February 1898 and arrived in New York harbor at approximately the same time on 27 February 1898 as did the *Manitoban*. (Maritime Reg. N.Y., 1898.) The sad song was of the accident at sea, 14 August 1888, when the *Geiser* (also of Thingvalla Line) on her way out of New York harbor bound for Copenhagen and the *Thingvalla* going westward to New York collided in haze or fog. The *Geiser* was sliced almost in two at the main mast, and 119 were lost at sea. (Hocking, 1969; Baker and Tryckare, 1965.) The *New York Sun* (28 February 1898) reported that the *Thingvalla* had departed from Trondheim, Norway, on 6 February, carrying about 500 reindeer, purchased by Max Jansen and shipped to the United States for speculation on the livestock market. Max Jansen at that time was under house arrest in Trondheim, Norway. (See above, Introduction, IX.)

Undoubtedly, the Brooklyn Bridge was that referred to; the impressive structure, with main span of 1595 feet and two side spans 930 feet each, was opened on 24 May 1883. (Shapiro, 1983.) The twenty-two-story building probably was the Municipal Building.

The *Manitoban* docked at the south side of Pier L, Harsimus Cove, Jersey City. Cattle yards were immediately adjacent. When the reindeer were taken off the ship, a writer from the

*Chicago Tribune* (1 March 1898, page 1) observed: "The *Manitoba* [*sic*] was tied up next [to] a cattle barge at the Central Stock Yards dock, and the reindeer were led from the ship to the barge and thence to the dock, past a crowd of a thousand or more curious persons, who cheered lustily whenever a reindeer ran off, dragging a bundle of fur with a Lapp or a Finn inside after it."

William Kjellmann also traveled with the trains. The presence of Lieutenant Devore in Europe as representative of the War Department may have been sufficient reminder to Dr. Jackson that the mission was in actuality a military operation, in the organization of which Kjellmann, an employee of the Department of the Interior, had no status. On 14 January 1898, in a special request to the War Department, Jackson pointed out the need for Kjellmann's service on the transcontinental trains and on the overland journey in Alaska. (War Dept., AGO File 67221, Nat. Arch.) Kjellmann was given permission to accompany the reindeer, with authority to act if necessary with regard to their welfare. Kjellmann did not well tolerate any ineptness in others' handling of the deer, however eagerly helpful in intent the military men may have been, and almost predictable was some friction between him and some of the army officers, whose methods of livestock management were those applicable to horses and cattle. Jackson asked General Alger to allow great discretion to Kjellmann, and evidently the secretary of war was happy to comply, for he ordered General Merriam, in command of the Department of the Columbia (which included Washington and Alaska), to see that his officers cooperated with Kjellmann. Even so, in Seattle when the trains arrived and arrangements so carefully laid began to fall apart and the deer were suffering and dying, Kjellmann, distraught, was not able to communicate with certain officers and went off in anger so that Jackson had to go and find him and try to persuade him to return to work, which of course Kjellmann did. But on the trains all went well.

**7 March.** Great Northern Railroad lines went to the long trestlework over the coastal lowlands just east of Everett, Washington.

Lieutenant Devore's assignment with the expedition terminated a few days after the trains arrived in Seattle, and as ordered he returned then to headquarters in Washington, D.C. Of the $60,000 for which he was responsible as disbursement officer of the expedition in Europe (plus $97.20 he had added to that amount through "sales"), Devore had paid out $39,657.12 (for ship's charter; passage and provisions for personnel; care of deer by Allan Line; purchase of deer, equipment, and fodder; and miscellaneous expenses); the bank of Seligman Brothers in London was holding $8394.71 subject to his draft on behalf of the War Department, and $364.98 was on deposit in Christiania in the name of the American consul, Department of State. The remaining $11,680.40 Devore turned back into the U.S. Treasury. (Alger, 1899, pages 40–41.)

**9 March.** In addition to Hedley Redmyer (spelled Redmyre in Carl Johan's journal), a second interpreter, Regnor Dahl, was hired by the War Department; monthly salary of Redmyer was $75, Dahl, $50, plus rations for each. Mrs. Dahl was hired as matron ($50 per month) for duty at Fort Townsend. (Letter from Jackson to the Secretary of War, 15 November 1898. War Dept., AGO File 67221, Nat. Arch.)

When possible, for uniformity of spelling, names of the people from Kautokeino (Finnmark) are given according to the Norwegian author Ørnulv Vorren. (Vorren, 1994.)

**13 March.** The episode was described in the *Seattle Post-Intelligencer*, 12 March, p. 8. In Washington, D.C., Dr. Jackson received a telegram from Kjellmann, asking for additional troops, as the visiting crowds could not be controlled by the few men present. (War Dept., AGO File 67221, Nat. Arch.)

**15 March.** Funeral services for little Mattis Nilsen Sara, three years old, son of Nils Person Sara and Inger Maria Martensdatter, were held in Seattle at the Bonney and Stewart Chapel, and burial was in Lakeview Cemetery.

**16 March.** The reindeer were to have departed at once for Alaska, but instead were held for nine days at Woodland Park, where (to save the short supply of reindeer moss for the trip north) they were fed fresh grass and hay, a highly inadequate diet. Perhaps a significant amount of moss had been lost in unloading from the *Manitoban* and the trains. Twelve deer died at and shortly after leaving the park, and others were markedly weakened. The *Manitoban* had been used in transporting domestic livestock in Europe and perhaps was not uncontaminated, but in the reindeer at Woodland Park and the Haines area, signs of disease other than malnutrition and starvation were not observed. Kjellmann felt it outrageous that the animals had not been sent north, but he could do nothing more than inform Jackson (by telegram on 10 March) that management of the deer was deficient and that few would survive for use in Alaska. No arrangements had been made for transportation to Alaska, or communications had been hopelessly snarled; no one seemed to know. Also, it was said that the army would keep the animals at Woodland Park for a month. When Lieutenant Devore was ordered back to headquarters in Washington, D.C., Captain William W. Robinson, Quartermaster Department, U.S. Army, became the officer responsible for the reindeer in Seattle, and he found himself in a difficult position while decisions concerning the animals were changed three or four times, under a new mandate.

On about 1 March, the secretary of war had cancelled the relief mission after receiving information from Major L. H. Rucker (Fourth Cavalry; stationed at Dyea and regional manager for the relief mission) that because so many of the destitute gold miners had left the Klondike for areas where food was available (Dyea, Skagway, and down the Yukon River to other villages), a critical shortage of supplies no longer existed near Dawson; alternative plans were discussed. For use in May and June by expeditions (to which the three scouts, Gumaer, Kelly, and Rafferty, eventually were assigned) under command of Capt. W. R. Abercrombie, Second Infantry, to explore the Copper River country, and Capt. E. F. Glenn, Twenty-fifth Infantry, exploring the Cook Inlet area, two hundred reindeer were to be retained in Seattle. The remaining deer were to go to Dawson or Circle City with Capt. Bogardus Eldridge, Fourteenth Infantry, and Kjellmann's men, traveling through the eastern margins of the St. Elias Mountains by Chilkat Pass, a pathway to and from the upper Yukon country used for hundreds of years by the local inhabitants but not well known to newcomers to Alaska. (U.S. Congress, Senate Mil. Affairs Com., 1900.) The trader Jack Dalton had driven a number of cattle to Dawson over that pass, probably after the Chilkat people had shown the way; the route was known to army personnel as the Dalton Trail. (Brooks, 1906.) Finally, the Department of War sent all the deer to Haines Mission (on the eastern arm of Lynn Canal); some were to be held there for Abercrombie and Glenn, the rest to go to the interior via Chilkat Pass. Carl Johan did not mention that the Norwegians were timely informed of the changed objectives of the expedition.

**17 March.** The ship was the famous old sailing vessel *Seminole*, 196 feet, built in Connecticut in 1865 as a half-clipper, and rated then amongst the best of her class. She was fast and dependable even though well worn as she was by 1898, and had been altered and rigged as a barque. The steam tug *Sea Lion* towed her to Haines Mission, perhaps because of doubts about her seaworthiness, but the *Seminole* sailed later to Australia and there, under tow by a steamer that itself had difficulties, was freed from the lines and arrived in port well before the tow vessel came in. (Matthews, 1930.) Going to Haines Mission aboard the *Seminole* with the starving and exhausted reindeer were fifty-seven of the men from Norway, along with Captain David L.

Brainard, Army Quartermaster Department, and William Kjellmann, who was again co-leader; forty people from Norway remained at Fort Townsend Barracks in Washington State. On 15 March, the *Seattle Daily Times* (p. 9) announced the departure: "The *Seminole* is receiving the Government reindeer and their unwashed Lapland herders." Probably with nothing more than buckets of water on board the *Manitoban*, on the transcontinental trains, and in the quarters at Woodland Park, the travelers undoubtedly found themselves soiled and worn; they must have missed sorely the steam bath house (saw'dne) and other means for personal hygiene that they enjoyed at home.

**21 March.** North of Queen Charlotte Sound four days out of Seattle, the *Seminole* was probably near Bella Bella, or Tolmie Channel. Charlotte Harbor is not listed in gazeteers as a town.

**24 March.** The customs station was then on Mary Island (55° 05' N, 131° 12' W), east of Annette Island. Jackson's ship passed the *Seminole* on 25 March, probably just north of Clarence Strait, near Kupreanof Island; Jackson left no precise record of his travel to Haines Mission.

**27 March.** The soldiers, led by Capt. Eldridge, were troops of the Fourteenth Infantry (in command of all was Col. Thomas M. Anderson). Eldridge and his men were to go over Chilkat Pass with Kjellmann, reindeer, and the company of men from Norway.

**28 March.** The War Department had planned months in advance for arrival of the deer at Pyramid Harbor, and Jackson had reconfirmed with General Merriam (commanding the Department of the Columbia) that a barge or other shallow-draft vessel would be present to transfer the animals to shore at once when the *Seminole* came in. But the plans had fallen apart. War with Spain was threatening, and General Merriam was obligated to consider the event of actual conflict; as well, relief for the Yukon was now unimportant. No orders from the general had arrived at Pyramid Harbor or at Haines Mission; no lighters were present in the harbor when the *Seminole* anchored offshore. Captain Brainard found a small boat and hurried to Dyea, at the end of Taiya Inlet, the eastern arm of Lynn Canal, about twenty miles (one way) from Haines Mission if the *Seminole* had rested in Chilkoot Inlet and approximately twice that far if Brainard had started out from Pyramid Harbor (Chilkat Inlet); he returned with a barge by which the reindeer were put ashore. From Woodland Park, they had been en route to Alaska for twelve days. Kjellmann (and Jackson) had planned that the overland party would depart Haines Mission immediately upon arrival there. Not only did they have to spend time searching for a barge, but Kjellmann found that all of the reindeer moss on the *Seminole* had been used en route. As well, the supplies needed for the men on the cross-country trek had not been sent; General Merriam's order that equipment and provisions were to be at hand when the reindeer reached Pyramid Harbor (or Haines Mission) had gone astray, forwarded to Dyea by letter from Skagway, less than ten miles but delayed almost a week. As a result, Kjellmann and his men waited another seven days for provisions to arrive (probably some from Seattle); although CJS mentioned (3 April) the "terrible rainy weather," evidently all of the party from Norway would have been willing, and able, to start off over the Chilkat Pass trail without rain clothes for the coastal area and with only makeshift shelters, since they were deeply concerned about the condition of the deer, but some supplies of food for sixty-five men (estimated number) were necessary.

**2 April.** Later, Carl Johan grouched (added to his journal) that Jackson was leaving "in order to enjoy good times, which we as yet knew nothing about. We learned to know Dr. Jackson more and more, but we still believed what he said."

**5 April.** Chilkot (Chilkat), some miles south of Haines Mission and abandoned in about 1910, was later reoccupied. Klukwan Village was also known as Chilkat. (Orth, 1971.)

**6 April.** The great avalanche occurred during the afternoon of 3 April 1898. Many of those caught in the snow were able somehow to free themselves. As often is the case with disasters, estimates of the number of victims were overstated; the *Seattle Post-Intelligencer* (9 April 1898) gave the figure as 150 but on 16 April reported that fifty-four bodies had been recovered and seven people were missing, as certified by Dr. Cleveland and U.S. Commissioner J. U. Smith.

**8 April.** The company went from Haines Mission along the braided Chilkat River to a point northwest of Klukwan Village where the Klehini River joins the Chilkat. There the men crossed over a relatively narrow ford and made their way northwest along the Klehini River. The route is essentially that of one of the mountain passageways (Chilkat Pass) into the interior used traditionally by the Chilkat people, and of the present Haines Cut-off Highway. (Chilkoot Pass was another in the old trail network between the coastal and inland tribes.) (Norris, 1996.)

The Norwegians probably were not aware of the hazard in walking through the abundant, and sometimes dense and high growths (to eight or nine feet) of Devil's Club *(Oplopanax horridum)*; clothing may not protect the unwary from its thorns, which can bury themselves in one's flesh and if not removed (removal is in itself difficult) cause painful local inflammation.

**9 April.** Klukwan Village: approximately twenty-five miles northwest of Haines Mission. One of the Chilkat men accompanied the expedition along the Klehini River. (Jackson, 1898b.)

**14 April.** Cornmeal or oatmeal and bacon, probably.

**27 April.** Redmyer's camp was on the Klehini River, northwest of Klukwan village.

**6 May.** At approximate lat. 59° 25' N, near the junction of Jarvis Creek, the Klehini River narrowed; flowing from NNW, it coursed through alpine meadows, having formed with its tributaries a natural passageway to an elevation of about 7300 feet, some miles to the north of the area where the two groups of men separated (approximately fifty miles from Haines Mission); from that point (the "north summit"), fifteen men of the expedition (including Redmyer) and 185 reindeer were to continue along the Chilkat Trail to Chilkat Pass (which marks the division of the coastal and Yukon drainage systems) into the interior of Alaska, to Circle City on the Yukon River. (Jackson, 1898b, 1900; War Dept., AGO File 67221, Nat. Arch.) Redmyer's account of the difficult trek was published in *Skandinaven*, 31 July 1899; a portion of the account was translated from Norwegian language to English, by Sverre Arestad. (Arestad, 1951.)

The troops of the 14th Infantry did not accompany the group going to Dawson. In mid-April, Jackson returned from Haines Mission to Port Townsend on the steamer *Queen*. (cf. p. 44, Jackson, 1898b; p. 18, Jackson, 1900; p. 6, *Seattle Post-Intelligencer*, 16 April 1898.)

**7 May.** The camp was on the western shore of the Chilkat River, which by early May probably was like the Klehini—becoming wide and swift with waters of the spring melt.

**8 May.** Peder Berg lost his pack in the Klehini River. With Kjellmann, forty-three men of the expedition returned to Haines Mission, according to Jackson. (p. 44, Jackson, 1898b.)

**9 May.** On the Chilkat Peninsula, near Haines Mission, forty-three reindeer remained (Jackson, 1898b); Captain Eldridge reported that sixteen deer were on the peninsula (ltr. dated 8 May to Adjutant General, Dept. of the Columbia) (War Dept., AGO File 67221, Nat. Arch.) Whatever the number, they were not used by the military or geological survey parties for which they were reserved, and their eventual disposition is not known; perhaps it was to some of those animals that a reporter from the local newspaper (*Seattle Post-Intelligencer*, 29 May 1898, p. 16) referred with some acrimony, in citing slaughter of reindeer by the local Chilkat hunters, probably in an undeserved accusation against those people.

On 25 April 1898 Congress declared war with Spain, retroactive to 21 April, after Spain's declaration of war against the U.S. on 24 April 1898. Since 15 February when the *Maine* sank in the harbor at Havana (Cuba), discordant communication had passed between the U.S. and Spain. Weeks prior to open hostility, the front page of the *Seattle Daily Times* (9 March) printed two two-inch columnar headlines: "WAR!" in one column, "GOLD!" in another to the right. Preparations for war, however at first tentative, had a definite effect on the Yukon Relief Expedition (which after early March 1898 might have been perceived as an adjunct of the long established reindeer industry and educational programs, now to be expanded through experimental use of deer as vehicles in exploration of Alaska and development there of reliable federal mail service), for the Department of War needed to give first attention to what was perceived as the nation's primary need under the circumstances—the defense of American property and rights in the Caribbean area.

On 4 April, the reindeer were turned over to the Department of the Interior, to be the responsibility of Dr. Jackson.

Except for the sledges, most of which were retained by the army, the reindeer-driving equipment purchased in Norway, and all supplies (camping gear, clothing, and food) allocated by the army quartermaster to the relief mission were stored in the building constructed by the men of the expedition. (Jackson, report of 15 Nov 1898. War Dept., AGO File 67221, Nat. Arch.)

**11 May.** Two companies of the 14th Infantry (led by Captains Yeatman and Eldridge) remained at Dyea and at Wrangell, under command of Anderson. (Merriam, 1898.)

**15 May.** Carl Johan did not write the name of the ship; Wilhelm Basi (p. 8, Diary; Finn. Amer. Soc., 1971) recorded that the vessel was the "Sytiof." Since no vessel so named ever existed, we thought at first that the ship was the Cunard steamer *Scythia*, inasmuch as she was listed as inbound to Victoria and Seattle throughout spring 1898, when she had been assigned to runs in the Pacific Northwest, but *Lloyd's List* makes clear that those schedules were not carried out; *Scythia* was at port in Liverpool from 1 May to 1 September 1898, and she never sailed to or from Seattle or Victoria. Probably the ship that stopped at Haines Mission to take aboard the people of the expedition was the steamship *City of Seattle* (Washington and Alaska Steamship Company); during May and June 1898 she called at ports along the Lynn Canal. She was rated as a first-class vessel, and could carry 600 passengers. On 19 May, en route from Dyea and other settlements, she arrived in Seattle a little after midnight. Wilhelm Basi may have heard her name as "City of" and wrote it as it sounded to him. At the time, he had very little knowledge of English.

In late May–early June, the supplies of food not used by the expedition were to be sold in public auction at Dyea (by order of Brig. Gen. H. C. Merriam, Commanding General, Department of the Columbia). On 6 June, Jackson in a letter to the assistant secretary of war suggested that the excess butter and bacon be turned over to the Interior Department for "use of the Lapps." The bacon, he explained, was not in very good condition, thus not saleable, and the butter also was stale. No one bid on either bacon or butter at the auction. Jackson also informed Assistant Secretary of War Meiklejohn that the merchants of Dyea considered themselves grievously injured commercially when the federal government offered the relief goods for sale, thereby competing with them for customers. Assistant Secretary Meiklejohn requested advisement from the adjutant general's office as to the legality of such a transfer as Dr. Jackson suggested. On 23 June, Judge Advocate-General G. Norman Lieber gave his opinion that since the reindeer and food supplies for attendants were purchased for the Yukon Relief Expedition, and since the reindeer had been turned over to the Department of Interior, the Department of War was authorized to transfer bacon and butter. (Letter of Jackson to

Assistant Secretary of War Meiklejohn, 6 June 1898 and endorsements et sqq. War Dept., AGO File 67221, Nat. Arch.)

**9 June.** The vessel was the schooner *Louise J. Kenney*, owned by Seattle Hardware Co. Transportation costs on domestic (U.S.) vessels for the expedition's personnel and supplies, from Seattle to and from Alaska ports, came to a total of approximately $20,700. (Alger, 1899, pages 45–47.)

**18 June.** The certificates were declarations of intent to become U.S. citizens.

**21 June.** Jackson negotiated with Seattle Hardware Company to ensure that the only passengers on the *Louise J. Kenney* when she departed Port Townsend for St. Michael would be the people of the reindeer expedition. Until he received word from the men at Port Townsend, he did not know that thirty-seven (thirty-five, according to Jackson) miners with all their equipment also had been taken aboard as passengers; the little vessel (96.8 feet, 155 tons), badly overcrowded, was in violation of certain federal regulations. He went at once (22 June) to Port Townsend and adjusted matters by removing from the *Louise J. Kenney* thirty-six of his people. That Dahl had informed Jackson of the situation is a possibility, although the author seemed most reluctant to credit Dahl with that judicious act. Whoever informed Jackson took care to inform the newspapers as well, but the report, if a strategy, rather miscarried. "Lapps Rise in Revolt," blared the *Seattle Post-Intelligencer*..."after enjoying several months of idleness and luxury at the expense of the government...." (*Seattle Post-Intelligencer*, 22 June 1898, p. 3.) The people from Finnmark, enveloped as some (including Carl Johan) were in their own prejudices, faced in their turn the derision of ethnic and national bias.

**22 June.** The schooner *Louise J. Kenney* sailed with thirty-five of Jackson's company on board, along with the miners.

Eight of the fifteen men who had been with Redmyer on the Chilkat Trail returned to Seattle from Haines Mission aboard the steamship *Farallon* (West Coast Steam Navigation Company), arriving on 21 June.

**23 June.** The steam schooner *Navarro*, 232 tons, was operated by Seattle-Skagway Steamship Company.

**25 June.** Sailing north in the Inside Passage, Captain C. E. Allen in command, the two-masted steam schooner *Del Norte* (122.6 feet, 233 tons), was nearly overburdened; she was towing the river steamers *Milwaukie*, *Mildred*, and *Wilbur Crimmon*, loaded with coal for refueling en route as well as machinery and lumber and miscellaneous freight, mostly for the gold camps in the Klondike region, and a fourth vessel, the dredger barge *Wisconsin*, which was also carrying a load of coal and lumber. That the *Mildred* developed a disabling leak in relatively calm waters and Captain Allen had to leave her on the beach at Juneau may have been fortunate, for the remaining journey was through rough and open ocean in the Gulf of Alaska and Bering Sea. (Newell, 1966; also see *Seattle Post-Intelligencer*, 13 July 1898, p. 9.)

Jafet Lindeberg stated that he had resigned from the government service upon his return to Haines (from the trek on the Chilkat Trail, April–May), but shortly afterwards, in Seattle, rejoined the reindeer expedition, with the intent to travel to the Klondike via the Yukon River to search for gold. (Carlisle, 1964.)

**27 June.** The steam schooner *Humboldt*, originally intended as a lumber vessel and half-built when the need intensified in 1897–98 for accommodations for prospectors and commercial agents going to gold fields in Alaska, was converted to a passenger ship. (Newell and Williamson, 1959.) She departed Seattle for St. Michael at about the same time as did *Navarro*;

Kjellmann was listed as a passenger and probably Jackson also was aboard. (*Seattle Post-Intelligencer*, 29 June 1898, p. 12.)

**25 July.** Carl Johan and his comrades spent almost a full month on the *Navarro* en route to St. Michael. Jackson (1898b) in error gave date of arrival of the *Navarro* as 27 July.

**26 July.** Port of St. Michael: 63° 29' N, 162° 02' W (on St. Michael Island). In 1833, the Russians established a fort at the locale, near villages of the Ungalaqligmiut (Malimiut). (Michael, 1967; Ray, 1966.) From 1897–1922, the United States Army maintained a military post there (Fort St. Michael). The commander of the post (Lt. Walker) reported to the adjutant general that the harbor, with shallow water and many rocks, was not a good one; ocean-going vessels of deep draft often had to anchor five to six miles out, and passengers and freight off the ships were rowed ashore in small boats. (Walker, 1899.)

Four men from the reindeer expedition were aboard the *Del Norte*.

**27 July.** Cloudberry: *Rubus chamaemorus*. The priest was Reverend William Furman Doty, the new teacher-minister assigned to the Presbyterian Mission school on St. Lawrence Island. Only two months earlier, the former teachers, Vene C. Gambell and Nellie F. Gambell, returning with their young daughter Margaret to duty on the island, had taken passage on the clipper whaler *Jane Gray* in Seattle; on 22 May they were lost at sea, with thirty-three others, when the vessel sank ninety miles off Cape Flattery. Vene and Nellie Gambell were the first teachers at the mission school in the village of Sivuqaq on St. Lawrence Island, living there half in fear evidently, their quarters ringed with barbed wire (Kelly, pp. 78–87 in Jackson, 1898a); in 1898 that village was given the name Gambell in their honor. (Hughes, 1960.)

**28 July.** The crackers in which some sort of fungus had found a life probably were among the inferior items (along with the bacon and butter) declined by bidders at the government's auction in Dyea in late May–early June.

About $13,000 in provisions from quartermaster's supplies were allocated for the people of the expedition; that amount included food items intended for the Yukon and turned over by the War Department to the Department of the Interior. (Alger, 1899, page 37.)

**31 July.** The schooner *Louise J. Kenney* arrived at Unalakleet on 29 July, left off thirty-five people of the reindeer expedition, and sailed northward to Golovnin Bay and on through Bering Strait.

Dr. Albert N. Kittilsen was an employee of the Bureau of Education, Department of the Interior, serving at Teller Reindeer Station, Port Clarence, as physician (1896–1897) and in the absence of William Kjellmann, as manager of the station (1897–1898) before he was transferred to the unit at Unalakleet as physician and manager. Kittilsen resigned that position immediately after Kjellmann's return in July 1898 and left for Golovin and Council City (a new settlement, established by gold prospectors) in order to concentrate on gold mining. (Jackson, 1898b; Kittilsen, 1900.) Actually, he had entered the field two months earlier, in April at Council City, when with P. H. Anderson, H. L. Blake, N. O. Huldtberg, D. B. Libby, L. F. Melsing, A. P. Mordaunt, and J. S. Tornensis, he was involved in organizing the Discovery (on Melsing Creek, near the settlement) and the Council City gold mining districts (cf. Carlson, 1946).

Rev. Axel E. Karlson was the clergyman at the church supported by the Swedish Evangelical Mission Covenant at Unalakleet and superintendent of the Mission's school. Other teachers were August Anderson, David Johnson, Malvina Johnson, Hanna Swenson Karlson, and Alice Omekejook. (Jackson, 1899.)

**1 August.** According to Vorren (1994), Frederik Larsen was from Kautokeino.

**3 August.** Edwin Engelstadt was manager of the Alaska Commercial Company's store at Unalakleet. (Carlson, 1946.)

**9 August.** Silver salmon *(Oncorhynchus ķisutch)* migrate in great numbers annually in August in the Unalakleet River.

**18 August.** Jackson returned from St. Lawrence Bay (Zaliv Lavrentiia) in Siberia to Port Clarence (Teller Reindeer Station) with 161 reindeer, 100 of which had been acquired (by barter of trade goods since United States' money was not useful in Siberia) by agents of the Bureau of Education, John W. Kelly, A. St. Leger, and Conrad Seim. According to Jackson, the three men had felt their lives endangered by the local people and had abandoned the U.S. government's purchasing station on 3 July, leaving aboard a whaling ship, after which "the natives seemed to have appropriated whatever they could lay their hands on, including the deer in the herd." (Jackson, 1898b.) The people living near St. Lawrence Bay were the indigenous Chukchi. (Levina and Potapova, 1961.) Seim may have felt in danger for his life, but he accepted employment with the J. S. Kimball Agency of San Francisco and was eager to return to Siberia in 1899 to purchase deer under a contract between Kimball and the Department of Education. (Jackson, 1900.) Carl Johan's report was third hand, but Kjellmann presumably told the story as he had heard it from Jackson. The three purchasing agents were never charged with misuse of government property.

Torvald Kjellmann was assistant superintendent of herds at Teller Reindeer Station.

**29–31 August.** Jackson had suggested often in communications to the federal departments that reindeer teams would be of use in transporting the mails during winter in interior Alaska; in spring 1898, his proposal was accepted and applied to the mail routes along the Yukon River, on which, also, some of the men from Norway at Eaton Station were to be assigned. Since the time Alaska became a part of the United States, mail delivery had been a problem in the district and functioned not at all, or haphazardly, or with great delays, and the governors (beginning in 1884) in annual reports to the president never failed to request improvement. To handle the mails for settlements along the coast in southeastern Alaska and northwest to Unalaska-Dutch Harbor, postal contracts for monthly schedules were arranged with transportation companies and other enterprises, among which were Alaska Commercial, North American Commercial, Pacific Coast Steamship, West Coast Steam Navigation, and Pacific Steam Whaling companies, but service was not consistent even during summer. Wells Fargo was successfully carrying freight inland and may have handled some of the overland mail to the Klondike, but everywhere in the interior, delivery and pick up were irregular. (Gruening, 1954; Hungerford, 1949; U.S. Congress, H. ExDoc. 1, 1884–94, H. Doc. 5, 1895 et sqq.; Ray, 1899a.) With the gold discovery in Yukon District and growth there and in adjacent Alaska of a population of U.S. citizens well able to express grievances, improvement in the mail service became necessary regardless of harshness of weather and 1500-mile mail routes. In summer 1898, the Post Office Department sent to Alaska two postal inspectors empowered to appoint postmasters and approve their bonds, let mail contracts, appoint clerks, and thus in effect to establish post offices in the course of their evaluation of conditions in southeastern and interior Alaska from Dyea to Circle City and along the Yukon River to St. Michael. (U.S. Post Office Dept., Ann. Repts., 1898, 1899.) That year, the Post Office Department was willing to experiment with different vehicles for mail carrying; the *Seattle Post-Intelligencer* (p. 6, 7 April 1898) reported that the department used reindeer for that purpose for the first time in Alaska, when Postal Inspector John P. Clum made a four-mile trial run with mail between Haines Mission and Chilkat, with apparent approval of the means. Following that test, Clum recommended that local postmasters be authorized to employ special carriers. (Dunham, 1898.) Contracts were drawn up by the postmaster-general, including one with Percival C. Richardson for delivery of mail using reindeer and/or dog teams on routes to and from villages along the Yukon River. Richardson planned to subcontract; with him, Wilhelm Kjellmann had agreed to be responsible for the mails in the western Yukon area from St. Michael to Weare (an outpost of the North American Transportation and Trading Company near the junction of the Tanana River with the Yukon, about a half mile west of the present-day town of Tanana), but the arrangement was not carried through. Among those at Eaton Station, Carl Johan was one chosen to participate (assigned to the military contingent), and his disappointment is indicated, but he probably was critical of the wrong party. Kjellmann (with bond secured by Sheldon Jackson) was later in charge of mail service between St. Michael and settlements on Kotzebue Sound (Route 78110, 1240 miles round trip), under subcontract with the North American Transportation and Trading Company, and Kjellmann and Edwin Engelstadt together had the contract with the Post Office Department for service between Eaton Station and Nome. (Jackson, 1900, 1901.)

**2 September.** Richardson (who evidently had traveled to St. Michael in order to arrange details regarding the subcontract) wrote from St. Michael in late September 1898 that since he had been unable to contact either Jackson or Kjellmann at Eaton Station, he had been obliged to sublet the mail contract to R. T. Lyng (of the Alaska Commercial Company and also postmaster) at St. Michael. (Jackson, 1900.) Lyng must have been an able organizer; the mails were successfully carried twice every month throughout the winter between St. Michael and Weare. Ice forming on the river in autumn and breakup in spring caused the only interruptions on that mail route for the year. (U.S. Post Office Dept., Ann. Repts., 1900.) A contract for the mails could be a valuable paper for the primary agent. Richardson received about $23,000 for the St. Michael-Weare route, and he offered Kjellmann a subcontract for mail service two times monthly from October through May between St. Michael and Weare with remuneration of $12,000 per eight-month period; for those months also, he offered Kjellmann a second subcontract for the route between Weare and Circle City at a salary of $15,000. (U.S. Post Office Dept., Ann. Repts., 1898; Jackson, 1898b, 1900.) Kjellmann probably received more money through making his own arrangements for other routes that were closer to his base (Eaton Station), but he never informed his men about the change of plan.

On 19 August 1898, the schooner *Louise J. Kenney*, after dropping off thirty-five people of the reindeer expedition at Unalakleet and continuing on her way to deliver the annual supplies to mission schools, was caught in shoreline breakers and was swept onto the shoals near Cooper's Station, eight miles east of Point Hope. Captain Francis Tuttle, commanding the Revenue Cutter *Bear* on her return from Point Barrow after rescue of the whaling crews stranded by ice during the previous season, learned at Point Hope of the schooner's situation and at once was underway to assist. The *Bear* reached the shoals on 20 August; the *Louise J. Kenney* was broadside on the beach, full of water, and in the heavy surf the cutter could do no more than take aboard the schooner's master and crew. No lives were lost, but her cargo, the year's supplies for the mission station of the Friends (Quakers) at Cape Blossom (Kotzebue Sound), could not be retrieved. (Tuttle, 1899.) The vessel herself was not lost; in September, she was pulled or floated off the beach and taken to Seattle (arriving on 17 October 1898); she later sailed in the fishing fleet of James Tyson in California. (*Seattle Daily Times*, 19 October 1898; U.S. Treasury Dept., 1899a; Newell, 1966.)

**4 September.** Jackson included a logbook, attributed to William Kjellmann, of events and weather at Eaton Station. (Jackson, 1900.) Since Kjellmann frequently was absent from the station, one assumes that others kept the record, but no acknowledgement was given.

**9 September.** Dr. Francis H. Gambell, employed (beginning in summer, 1898) as physician at Eaton Station after A. N. Kittilsen resigned that post, was the brother of Vene Gambell (lost at sea with his wife and daughter in May, on the *Jane Gray*).

**10 September.** Dr. Gambell's evening school provided Carl Johan's first formal lessons in the English language.

**17 September.** Some of the food supplies at Eaton Station were from the stocks purchased by the Department of War and turned over to the Department of the Interior (Bureau of Education) for use by personnel of the Yukon Relief Expedition.

**24 September.** *Nafta* was the brand or trade name of a petroleum compound, aromatic and pleasant to the taste, evidently widely used to relieve symptoms of colds and respiratory infections. Dr. Gambell probably did not have that product in his dispensary as it was of the European market.

**28 October.** According to Jackson, Engelstadt and Kjellmann jointly owned the mail contract. Jackson provided surety for Kjellmann's bond. Later, Dr. Gambell (at Eaton Station) had the contract. (Jackson, 1900, 1901.)

**1 November.** The story of Jafet Lindeberg and his rich gold find has been told often in the literature. The account given by L. H. Carlson is a well-documented review of the events and people involved. (Carlson, 1946.)

**4 November.** That dietary deficiency might have caused problems is understandable, particularly if people had been relying on the preserved goods purchased by the Department of War for the relief expedition. Although fresh fish from the Unalakleet River and wild plants were available, the tinned foods, intended for short-term use on the expedition, may have ended up as the chief sustenance for some of the people at Eaton Station.

**14 November.** Downriver, the local people probably were catching saffron cod (tomcod) *(Eleginus gracilis)*. E. W. Nelson in ethnographic studies along the western Alaska coast observed (p. 175) "In autumn the tomcod are extremely abundant near St. Michael. At this season cold north winds generally blow and render it very uncomfortable to remain for hours in one position on the ice.... In November, soon after the ice is formed, a fisherman frequently catches 200 pounds of tomcod in a day, but from 10 to 40 pounds is the average result of a day's fishing." (Nelson, 1899.)

**15 November.** John L. Hagelin went to Golovnin Bay in spring 1898 as a coal miner employed by the Good Hope Mining Company; soon after arrival, he became interested in prospecting for gold. (Cf. Carlson, 1946.)

**22 November.** In Dawson, among patients at the Catholic Hospital, Dr. J. J. Chambers diagnosed more than twelve cases of scurvy in December 1897 and expected that several hundred people additionally would be afflicted by spring. (Wells, 1899; 1984.)

**24 November.** Thanksgiving Day 1898. The account about the findings at Cape Nome as told by the author, who heard about them directly from Lindeberg, is similar to that given by Carlson (1946). Lindeberg's description of the gold discovery was written some years later in a letter to Frank L. Hess of the U.S. Geological Survey (pp. 16–18 in Collier et al., 1908).

Nels O. Huldtberg had been sent as a missionary to the church at Golovin by the Swedish Evangelical Mission Covenant. In 1898 he was closely associated with some of those involved in the first gold finds and indeed may have recognized before anyone else the richness of a small stream later named (by Jafet Lindeberg) Anvil Creek, flowing from a highland on which a rock outcrop when seen from below resembled in shape a blacksmith's iron block, but that year Huldtberg did not lay claim to any sites. In succeeding months at Cape Nome, he made significant profits financially, but evidently he felt himself cheated of even greater wealth by his companions and brought suit against some of them, ostensibly partially to benefit his church; the court cases that arose (to continue for almost twenty years) were bitter and costly beyond money as various individuals sought portions of the fortune(s) taken from the streams near Anvil City. (Cf. Carlson, 1951.) Jafet Lindeberg later would be involved unavoidably in lawsuits instigated by claim jumpers, but that he had the means, and the foresight, to purchase and transport his own supplies was fortunate, as at least no other agency could claim in court to have supported him monetarily or in any way that would jeopardize his ownership of gold claims established in the course of the prospecting trips of autumn 1898. (Cf. Carlisle, 1964.)

The seven who went on the second trip to Cape Nome were John Brynteson, Albert N. Kittilsen, Eric O. Lindblom, Jafet Lindeberg, Gabriel W. Price, Johan S.Tornensis, and Konstantin Uparazuck.

The physician serving at Fort St. Michael was J. R. Gregory, evidently on army contract.

Peter H. Anderson, a colleague of Huldtberg, was one of the clergymen (Swedish Evangelical Mission Covenant church) at Golovin (sometimes called Chenik in 1898).

**30 November.** The mail was perhaps coming also from Weare, upstream on the Yukon River about 250 miles from Nulato.

**3 December.** In exchange for transportation by reindeer from Eaton Station to and from St. Michael, Unalakleet, Golovin, Cape Nome, and elsewhere, the business enterprises at St. Michael and Unalakleet traded supplies for the station's commissary; also, some drivers were paid hourly.

**4 December.** Probably, to Carl Johan, that Dahl went for his revolver meant certain death for Larsen.

**6 December.** Lieutenant Edgar S. Walker, Eighth Infantry Regiment, was commanding the military post at St. Michael (Fort St. Michael). He was shortly to be promoted in rank.

**14 December.** According to Kjellmann's logbook, Dr. and Mrs. Gregory and Mr. and Mrs. Hatch were taken by reindeer sledges to Unalakleet. (Kjellmann was not present at Eaton Station when that information, and other entries, were recorded.) E. T. Hatch was deputy collector of customs at St. Michael.

**18 December.** Gold had been reported from the Unalakleet River, upstream from Eaton Station, in fall 1898. (Schrader and Brooks, 1900.) Arthur E. Southward had been prospecting for gold in that area. (Carlson, 1947.)

**24 December.** Klemet Nilsen was forty-four years of age when he left Norway.

**26 December.** The men walked to Unalakleet and back to Eaton Station.

**27 December.** Malvina Johnson and Alice Omekejook were teachers at the mission school in Unalakleet.

**31 December.** The lack of marrow fat as observed in bones of some of the reindeer at Eaton Station would suggest that those deer were not healthy. That the animals were of Siberian stock probably was not relevant, but their condition might have been attributable to overuse, through which they became depleted nutritionally.

## 1899

**5–11 February.** John Dexter had established a trading post and general store at Golovin and other villages. (Dexter, 1900.)

**12–18 February.** Rev. Karlson had been working in Unalakleet since 1887. (Jackson, 1898.)

**19–25 February.** Second Lieutenant Oliver L. Spaulding, Jr., Third Artillery Regiment, was leading the military group of five men; the judge was Lennox B. Shephard. (War Dept., Ann. Repts., 1899.)

**5–11 March.** Leinan and Dahl were sent to Cape Nome by the North American Transportation and Trading Company.

**19–25 March.** The names of the Iñupiaq and Yup'ik people were spelled in Norwegian and English as they seemed to be according to European orthographies, which could represent only approximations of the sounds of the various languages spoken in Alaska.

**26 March–1 April.** The United States commissioner at St. Michael was L. B. Shephard.

Federal commissioners in Alaska were lower-court judges, appointed by the president with the Senate's approval, to exercise duties as imposed on justices of the peace according to the general laws of Oregon, including powers to grant writs of habeas corpus, to decide in matters of wills and of probate, to record deeds and transfer of title and other legal documents, and to act as demanded of notaries public in Oregon. The commissioners received a monthly salary of a thousand dollars, plus fees collected as allowable by Oregon's laws. (Alaska Organic Act, 17 May 1884, 23 Stat. 24.) Shephard was also the agent of the North American Transportation and Trading Company at St. Michael. The claim jumpers taken under arrest to St. Michael were Louis F. Melsing and John Watterson. (U.S. Congress. Senate, Hearings, 1900.)

**9–15 April.** The Iñupiat observed that some of the Saami enjoyed alcoholic spirits.

**23–29 April.** Cho-kok (now known as Chiukak): a small village about ten miles west of Golovin.

**30 April–6 May.** Bonanza District was later included in the Council City District. Probably, one of the organizers of the mining districts around Council City was stationed at the cabin (or hut, as Carl Johan called it).

**7–13 May.** The small island (Cuba) where Carl Johan and Johan Petter staked their claims probably was named with thoughts of the war with Spain.

**14–20 May.** The information given by L. H. Carlson essentially is in agreement; in Carl Johan's journal, the data in one or two cases were not so complete, e.g., Discovery Claim on Anvil Creek was owned by John Brynteson, Eric O. Lindblom, and Jafet Lindeberg (who in September 1898 located and staked that field for themselves). Claim No. 11 Above Discovery was taken by Johan S. Tornensis for his relative, not named by Carlson but who may have been Mikkel Nakkila. The owner of Claim No. 12 Above was Jens C. Widstead (or Widsted; spelled in the journal as Hvidsted); Carlson indicated that Kittilsen used a power of attorney in Widstead's name to ensure the claim. Widstead had worked for the Bureau of Education as assistant superintendent (and superintendent in 1895–1896 while Kjellmann was absent) of the Teller Reindeer Station. (Carlson, 1946; Jackson, 1896.) Claim No. 12-b perhaps means No. 12 below Discovery. For Nakkila Gulch, Carlson stated that Claim No. 1 was that of Mikkel Nakkila.

At Cape Nome, gold flakes and nuggets had accumulated abundantly in several stream beds, including that of Anvil Creek, all claims along which were exceptionally productive. Claim No. 9 Above on Anvil Creek was found to be very rich. That site was staked by Gabriel W. Price using power of attorney of his brother R. L. Price, evidently with the understanding that R. L. Price was to be merely the trustee thereof, for the claim was located and documented with intent that Konstantin Uparazuck was the owner. That the indigenous Alaskans (who were not given full United States citizenship until 1924) (2 June 1924, 43 Stat. 253) were eligible to hold gold claims was in doubt amongst the claimants at Cape Nome in spring 1899, although that question had been resolved much earlier by the U.S. Supreme Court (see below). Uparazuck was well aware of the claim that had been made on his behalf but was not able to challenge the proceedings that followed, wherein his interest was repudiated, and Claim No. 9 Above was sold (evidently for twenty dollars) by G. W. Price to P. H. Anderson (a missionary of the Swedish Evangelical Mission church at Golovin). The church, Rev. Anderson, and Rev. Nels O. Huldtberg became involved in lawsuits, bitterly disputing among themselves for many years (from 1899 to 1919) over who could have the gold. (Carlson, 1951.)

Carl Johan in May 1898 noted without any doubt that Claim No. 9 Above was located for Konstantin [Uparazuck] and was held by Price [R. L.] as trustee.

Konstantin's relative Gabriel Adams (who died in 1900) was also associated with those who founded the Cape Nome Mining District, and was indicated as owner of one of the claims on Anvil Creek. A lawsuit (initiated by K. J. Hendrikson, a pastor of the Swedish Evangelical Mission Covenant church and with Uparazuck comanager of Adams' estate) charged Peter H. Anderson with deliberately cheating those rightful owners of their gold claims (cf. Thuland, 1950). In an out-of-court settlement in 1902, Uparazuck and the Adams' estate were awarded about $39,500 (including attorneys' fees). (Carlson, 1951.)

**21–27 May.** Not only was the work made difficult by the long winter's freezing, but the entire region was underlain to a great depth by permafrost. Arthur F. Wines (of the U.S. Census Bureau) remarked that "an underlying glacier" was present beneath the vegetation of the tundra; during the first geological surveys on the Seward Peninsula, Alfred H. Brooks and his coworkers noted that the mining operations in 1899 had "in no instance penetrated below" the layer of permanent ice. (Brooks et al., 1901; Wines, 1900.) The maximal depth of permafrost determined later by the U.S. Geological Survey was around six hundred feet on the peninsula. (In 1898, the governor of Alaska proposed that the peninsula be named in honor of William H Seward, but that designation was not immediately used.)

Two business establishments at St. Michael with sufficient resources to offer a prize of free passage were the Alaska Commercial Company and the North American Transportation and Trading Company.

**28 May–3 June.** The message probably was sent by Captain Edgar S. Walker, with acquiescence of Commissioner (Judge) Shephard.

**4–10 June.** The schooner may have been a vessel of the whaling fleet carrying passengers who paid a fare. From Anvil City (which was situated almost at the mouth of the Snake River), the Penny River was ten miles, and the Sinrock (Sinuk) River twenty-five miles, to the west.Westward from Golovnin Bay, as far as Port Clarence, the coast of the Seward Peninsula is relatively unbroken, with no natural indentation for harborage. Ocean-going vessels stopping at Anvil City (Nome) could approach land no closer than about a mile (or somewhat less, depending on draft) in the then-uncharted coastal waters of the Bering Sea, and passengers and freight were taken to the newly established settlement on small boats or barges. The mile or so from steamer to shore was hazardous, for winds could transform the shallow waters along the unprotected coast into turbulent and dangerous surf. Coast and Geodetic Chart No. 9302 (6 February 1971) shows the depth of the sea to be only two fathoms at Nome.

The claim jumpers were said to be men mainly from Council City who had been organized by J. L. Wilson and A. P. Mordaunt (the recorder for Eldorado Mining District) as a group ostensibly concerned with law and order in the Cape Nome District. According to information presented in 1900 to the U.S. Senate in hearings relating to civil government in Alaska, people from the eastern U.S. and Europe also were involved. (U.S. Senate, Hearings, 1900.)

**18–24 June.** Anvil Glacier was probably a small, deep snow field, slow to melt but advantageous since it must have fed water to the streams after the main breakup. Other than a few small bodies of snow or ice, evidently perpetual, found in the Kigluaik Mountains to the north, near Mt. Osborn, glaciers did not exist on the Seward Peninsula. (Brooks et al., 1901.)

**25 June–1 July.** Captain Walker sent Lieutenant Spaulding from St. Michael with a small group of enlisted men to try to ensure peace at Anvil City. Federal troops were the only governing force at Anvil City during most of the summer of 1899.

**9–15 July.** The meeting on 10 July 1899 was instigated by A. P. Mordaunt and J. L. Wilson, who with fifty members of the Law and Order League (aggressive claim jumpers mostly from Council City where the stream beds were not so rich in gold) had sent a signed petition to the national congress listing their grievances and asking for acceptance of their recommendations which included removal from office of Commissioner Shephard (who had sent two of their number to jail), Captain Walker, and Lieutenant Spaulding, and a relocation of the existing (and legally recorded) gold claims at Cape Nome Mining District. (U.S. Senate, Hearings, 1900.) About five hundred people crowded into a large tent shelter on 10 July, including a number of men from the Yukon Territory. John A. Dexter (from Golovin) evidently was present; he reported (in 1900) to the United States Senate that Lieutenant Spaulding had announced that anyone objecting to existing mining claims or wishing to change mining law would have to wait until the courts might decide in their favor "before they took possession of the claims which other people had located." (U.S. Senate, 20 April 1900.) Spaulding ordered his seven men to advance in a line, their rifles with bayonets fixed, and the crowd dispersed. Major General Adolphus W. Greeley (chief of the Army Signal Corps in Alaska) commented that those attending the meeting were "not the busy miners but...several hundred disappointed and idle gold seekers [there] for the understood purpose of vacating all miners' locations and throwing them open to the first—or in this case to the last—comers." (Greeley, 1909.) Greeley and others believed that Spaulding's quick action and later maintenance of a reasonable policy against which none dared openly to revolt forestalled a miners' war. (During the hearings of 1900, Senator Spooner of Wisconsin spoke to defend Lt. Spaulding against suggestions that he was corrupt, pointing out Spaulding's honorable service as an officer in the regular army and that he was a son of the assistant secretary of the treasury.)

Three days later, Captain Walker with Second Lieutenant Wallace C. Craigie, Second Infantry Regiment, and about twenty enlisted men arrived from St. Michael, at which time Walker reaffirmed Spaulding's orders of 10 July, that all disputes were to be settled before the civil courts. Lt. Craigie remained in Anvil City in charge of the military post there (later named Fort Davis). (Cf. Carlson, 1947a.)

Had the meeting of 10 July been successful in forcing cancellation of existing claims and a general relocation, at risk of loss were citizens and noncitizens alike who had staked and recorded for themselves, or using powers of attorney for others, the first tracts in the Cape Nome Mining District. At least twenty-five individuals stood to lose out, including several from the Yukon Relief Expedition (i.e., recent immigrants from Norway, e.g., Ole Baer,

Wilhelm Basi, Magnus Kjeldsberg, Thoralf Kjeldsberg, Jafet Lindeberg, Mikkel Nakkila, Olai Paulsen, Carl Johan, and Johan Sp. Tornensis), none of whom were U.S. citizens but all of whom had declared intent (in 1898 at Port Townsend and at St. Michael) to become citizens months prior to the gold discovery at Cape Nome. The remaining first-comers included men of various origins; some were naturalized citizens born in Scandinavia, the Iñupiat were Alaskans by birth, and others were Americans who had entered the district from the States (e.g., John A. Dexter, J. R. Gregory, R. T. Lyng, G. W. Price, R. L. Price). The late-comers Mordaunt and Wilson and others had been in a rage against the lot of them, perceiving unfairness about it all and a collusion that if they could not define they suspected, and they had taken matters upon themselves to jump original claims. It was only the beginning of much conflict, even involving some members of Congress, and occasional violence in the gold fields at Nome, attributable to the accursed hunger for gold: "Quid non mortalia pectora cogis, Auri sacra fames!" [Vergil, *Aeneid*, III.]

**16–22 July.** The ships went to the lee of Sledge Island, five miles off the coast and twenty-five miles west of Nome (Anvil City).

**23–29 July.** Jackson obtained eighty-three reindeer from Koriak herd owners in Siberia and the *Bear* (Lt. David H. Jarvis in command) delivered them to Port Clarence after stopping at St. Lawrence Island. At Cape Blossom, Kotzebue Sound, Lieutenant Jarvis took aboard more than eighty people who had been prospecting (with no great success) along the Kobuk and other rivers, and were suffering from nutritional and other sicknesses; in the Bering Sea on the way to Anvil City, the cutter passed by many small boats carrying miners from the Kobuk area en route to the Cape Nome gold fields. Before visiting Unalakleet, the *Bear* stopped at St. Michael (26 July). There, Jackson noted that "the trade...has greatly changed since a year ago. Then there was a rush of miners up the Yukon River; now very few are seeking passage up that river—but thousands are coming down; some to leave the country in disgust and others to try the Cape Nome mining district. The upriver business is now mostly freight. Owing to the decrease of the passenger traffic, many of the small river steamers are laid up. The harbor is full of them." (Jackson, 1900.)

**30 July–5 August.** After gold was discovered on the sea beach, hundreds of men turned their attention there; meetings of protest became less important, for no time was to be spent but on the beach, as with every scoop of sand, a prospector could almost count on being richer in money than he was before he began to dig. Other than purchasing a shovel, materials to construct a simple rocker to wash the beach sand, and (lacking copper plating) a blanket to place in the rocker to catch the metal, no capital investment was necessary; the beach by regulation was open to all. Along the beach tundra inland from high-tide line the Nome Mining and Development Company (owned by Kjellmann and four others) had extensive claims, and trouble began again when the firm attempted to limit entry by others, but was settled when, pursuant to land law, a shoreline strip sixty feet in width was recognized as reserved for a roadway, and prospecting there was open as well. (Dunham, 1900; Brooks et al., 1901.) Jackson described (26 July 1899) the beach at Anvil City: "...a conglomeration of tents, with half a dozen frame houses or shanties, and two or three iron warehouses in process of erection by the transportation and trading companies. The ocean front is staked out with claims for from 10 to 20 miles. We saw men panning out gold on the beach in front of the most densely populated part of the place. Some fine teams of horses were being used for hauling." (Jackson, 1900.) Schrader and Brooks (1900) stated that the beach gold had been first reported almost simultaneously by a soldier who had been digging a well and by prospectors, one of whom (John Hummel) had been panning sand by the sea because he was ill with scurvy and not able to walk into the interior country. (Others have given different versions of the finding.) The shortage of labor developed when men left the inland streams and went to the shore in order to work independently. Special Agent Arthur Frederick Wines, of the Twelfth U.S. Census, was present in Anvil City during summer and autumn 1899; his descriptions of the local conditions were specific and detailed. (Wines, 1900.)

**6–12 August.** Charles D. Lane, of the Wild Goose Mining Company.

**13–19 August.** Concerning rights of noncitizens (including the Iñupiat, although during the week it had been the Scandinavians who were accused of unlawful acts), the issue was whether an alien (i.e., noncitizen) could legally locate and hold a mining claim. The question seemed complex and the Alien Act of 1887 further obscured its resolution, but over that point the federal courts had indicated (based on findings of the U.S. Supreme Court, when John Marshall was Chief Justice, in employing the principle of inquest of office), in dealing with cases of alienage, that citizen or alien alike might make and hold a claim location, the competency of noncitizens therein not to be questioned except by the government; i.e., the right to defeat a title on grounds of alienage is reserved alone to the government. Numerous cases affirming the Supreme Court's rule could be cited as precedents (cf. Lindley, 1897); nonetheless, the question of noncitizens and gold claims became an important issue in the U.S. Senate in 1900. That year, to the Alaska Civil Government bill (S.B. 3419), Senator Henry Clay Hansbrough (North Dakota) and Senator Thomas H. Carter (Montana), both of whom by 1899 held shares in some gold mines at Cape Nome, asserting that the laws of Oregon that governed Alaska were not intended to include those relating to noncitizens and mining claims, attempted to attach an amendment. It seemed innocuous enough, but in effect, that change would redefine the mining law relative to aliens in the district, so as to abrogate their claims, thereby permitting restaking the gold fields and new ownership of the sites. Extensive debates took place in the Senate. The discussions on the floor brought support for the noncitizen miners at Anvil City, notably from Senators Henry M. Teller (Colorado) and William M. Stewart (Nevada), and on several occasions specifically dealt with the group of people who had come to the United States from Norway in 1898 by invitation of the United States government to serve on the Yukon Relief Expedition. Understanding the prejudgment inherent in the Hansbrough amendment and in opposition to its retrogressive, retroactive (ex post facto) substance, Senator Stewart on the Senate floor on 18 April 1900 attempted to rouse his colleagues, some of whom evidently were asleep or so much bored with the entire question that must have seemed so far away and unrelated to real problems that they appeared to be asleep; he apologized "for disturbing Senators in their repose to listen to this matter...but I wish attention for a minute or two before they again retire to their rest.... I hope more Norwegians and Laplanders will go there and explore that cold region where others can not live.... [They] are about as good bona fide residents as you ever have had up there." The Senate was fully awake during following days when Senators Stewart, Teller and others cited cases concerning noncitizens and mining claims, including the decision of the Supreme Court in 1886 in O'Reilly v. Campbell that mineral lands of the United States are open to citizens and to those who have made declaration of intent to become citizens of the United States (United States Repts., 1886). As originally written, the Hansbrough amendment to the Alaska Civil Government bill was not accepted; but a bitter fight concerning it took up most of the month of April on the floor of the Senate (cf. Bjork, 1985; Wickersham, 1938.) Among others, Sheldon Jackson, Albert N. Kittilsen, and Jafet Lindeberg were called to testify. (U.S. Senate, 1900; Jackson, 1900a; Kittilsen, 1900; Lindeberg, 1900.)

**20–26 August.** John Green Brady, appointed to the office of governor of Alaska by President McKinley in 1897, was a missionary of the Presbyterian Church; Sheldon Jackson had sponsored his first employment in Alaska, in 1878. Judge Charles S. Johnson was from the U.S. District Court in Sitka. Judge Johnson named individuals for positions as U.S. commissioner and U.S. marshall to serve at Anvil City (Nome), and advised the local people in their plans to form a city government.

**10–16 September.** Pursuant to the suggestions of Judge Johnson's court, the local election was held on 12 September, when city officers, including a mayor, a chief of police, and a health officer, were chosen by the electorate (1418 votes cast). (Carlson, 1947a; Wines, 1900.)

**17–23 September.** Charles D. Lane was purchasing claims, some from recent immigrants; although all who had sold their holdings had declared their intent to become citizens of the U.S., that the declarations had been lawfully executed was questioned, and Lane's ownership of the sites was thus subject to doubt. When the issue of alienage was under debate in the U.S. Senate a few months later (in 1900), Lane was called to testify, to defend his property against allegations of illegal procedure in the original staking of the claims he had acquired. Some of those had been located by use of powers of attorney, as well. (Lane, 1900.) The Hansbrough amendment would have affected negatively Lane's title to his property at Cape Nome.

The site of Anvil City had been changed almost wholly in a single year (1898–1899) into an area where traffic of people, livestock (dogs, horses, and reindeer), and wagons and sledges had worked a primitive, natural shoreline and tundra into a stretch of soil mixed with melt water and rainwater and effluents draining out next to tents and wooden structures. The town's water supply came from a small pool longside the Snake River, near its mouth, and from the river itself. Into the river the townspeople dumped their wastes, and from upstream at the gold camps flowed the substances deposited directly or drawn into the river and its tributaries. (Cf. Wines, 1900.) Sanitation was not a general concern at Anvil City, but all the elements (sewage, fecal contaminants in the water supply, lack of facilities for personal hygiene) were there to bring about an epidemic of typhoid fever, provided that the bacterium (*Salmonella typhi*) was present; obviously, it had been introduced by the newcomers, perhaps by a single carrier. (Cf. Rausch, 1974; Stürchler, 1988; Fortuine, 1992.) Conditions could only improve, and they slowly did in the following two years. (Cf. Carlson, 1947a.)

**24–30 September.** The steamers *Roanoke* and *Cleveland* arrived about as scheduled. Also the *Aloha*, the *Bertha*, the *Homer*, and the whaling bark *Alaska* eventually came and departed with passengers. (*Nome Gold Digger*, 25 October 1899.) The *Laurada* never made it to Anvil City.

Dr. H. Newton Kierulff as well was assigned to the U.S. Military Hospital, some time after Dr. James M. Miller departed. A second medical facility (the Hospice of St. Bernard) had been set up in the library building. In late October 1899, it was moved to the Ryan Building where in two wards and several private rooms about forty patients were to be accommodated. (*Nome Gold Digger*, 4 November 1899.) Earlier plans had called for a new building for the hospice, and Rev. Loyal L. Wirt, of the Presbyterian Church, sailed from Anvil City to Seattle in late August in order to obtain supplies for that purpose. He carried along funds donated by residents at the settlement and mining camps and collected further contributions in Seattle, so that altogether he was able to spend about $4,000 for medical supplies and equipment, furniture, and lumber. An unfortunate setback in the plans occurred in late September or early October, at the time Rev. Wirt returned on a steamer carrying the material, when the barge, loaded from the steamer (anchored out from shore about a half mile) with all that he had purchased, was swamped in a heavy sea as it attempted to get ashore at Anvil City. (Wirt, 1937.) Only a part of the supplies was saved, some evidently retrieved and kept by local people. In the *Nome Gold Digger*, 1 November 1899, on page 2, Rev. Wirt's supplication to the salvagers was published: "An Open Letter. Miners and Friends: The St. Bernard hospital outfit has been wrecked on these shores. In the name of the sick we appeal to the generosity of the brave men who rescued this wreckage from the waves to relinquish their rightful claims upon it. With the chairs, beds, books, furniture and supplies thus saved we shall fit up a temporary hospital in the warehouse so kindly placed at our disposal by the A. E. [Alaska Exploration] Co. The lumber and shingles are to be sold to the highest bidders and the funds thus realized shall be used in the purchase of another building to be erected at the earliest possible moment. Confidently relying upon the honor and cooperation of every man in this camp, believe me, Your brother and friend, Loyal L. Wirt. N.B. Deliver no goods to anyone except Raymond Robins, my chief of staff, who will give receipts and settle all claims in full." (*Nome Gold Digger*, 1 November 1899.)

**15–17 October.** The old but trusted steam schooner *Laurada* (built in England in 1864) had been on scheduled runs to Alaska for only about two years; earlier, she had worked out of eastern and southern ports of the United States, from where on one of her last journeys (1896) she had carried 321 Americans (former slaves) to Liberia, as new settlers hoping for better opportunities in Africa. (Kratzer, 1981.) On 12 September 1899 (Frank M.White, Captain), she sailed from Seattle, bound for Unalaska (Dutch Harbor) and Cape Nome with 68 people, 36 head of cattle, 130 sheep, and a full cargo of freight—lumber and supplies for the mining camps, and coal. Approximately twelve hours out of Seattle, in Active Passage, she took a sheer and was driven onto the beach just east of Helen Point; there she waited, with expectation to come off at high tide, which she did. A few hours later, just out of Dixon Entrance, Captain White, with his heavily laden ship running low in the sea, bucking the headwind, and with his concern for the livestock on deck, turned back into the Inside Passage to Metlakatla, the closest port, where part of the lumber and coal was put ashore. After departing from Dutch Harbor on 26 September, the *Laurada* encountered very heavy weather in the Bering Sea, straining and evidently tearing out some rivets in the iron plates of her forward hull; a leak developed, water gradually overwhelmed her pumps' capacity, and as her bow lowered hourly into the sea, Captain White headed the sinking vessel for the nearest land, which was St. George Island (Pribilof Islands). There, in Zapadni Bay (far from the usual routes of commercial ships) on 29 September, in order to keep from going under, the captain ran her aground. All the people aboard were able to reach shore. On 3 October the Revenue Cutter *Corwin,* sailing to the south of her usual course, by chance came upon the beached ship. She carried the passengers and most of the crew back to Dutch Harbor; nine men were left on St. George Island to watch over the wreck. (Bureau of Marine Inspection and Navigation. 1900.) The *Seattle Post-Intelligencer* (16 October 1899, p. 1) reported that the livestock were put at liberty onto the beach, and only one steer was lost when he ran into the sea instead of ashore, but Old Tom, one of the vessel's cats, refused to leave the ship. Captain White, who returned to the site of the wreck, informed the official inspectors that the ship and its cargo were a total loss. (Bureau of Marine Inspection and Navigation. 1900.) Concerning the *Laurada,* Schrader and Brooks (1900) stated that she was wrecked on St. Lawrence Island, as did Carlson (1947a), citing other sources; *Lloyd's Register* (*New York Maritime Register*, 18 October 1899) gave the locality as Zapodine Bay. Newell (1966) gave the site of the beached vessel as Zapotni Point, St. George Island. Carl Johan reported that she sank, which was also in error.

**18 October.** The *Portland* was a vessel still celebrated in 1899; she had carried in 1897 a ton or two of Yukon gold in a single voyage to Seattle from St. Michael and became known as a treasure ship, and a lucky one. She had not always had such good fortune; in the past (under her original name, *Haytien Republic*) she had been accused of smuggling guns in the Caribbean Sea in 1885 and Chinese people from Canada to the U.S. in the Puget Sound area in 1893. In 1897, renamed the *Portland*, she was chartered by the North American Transportation and Trading Company in Seattle for service to Alaska. She never left the territory; she was abandoned in 1910 after wrecking on an uncharted reef near Katalla, just north of Kayak Island in the Gulf of Alaska. (Newell and Williamson, 1959.)

•••

After completing his record of the events of 1898–1899, Carl Johan asked that his journal be read by four of his associates on the expedition. Wilhelm Basi, Per Johannessen Haetta, Isak Salomonsen Nakkila, and Ole Johannessen Pulk read the work and attested to its accuracy by their signatures, attached to the original document.

For more than a year (1898–1899), Carl Johan continued his service with the Department of War and the Department of the Interior. He returned to Alaska in 1900 but kept no record of his observations.

# Epilogue

Hedley E. Redmyer and six men of the Yukon Relief Expedition (Anders Aslaksen Baer, Klemet Persen Boine, Per Johannessen Haetta, Emil Kjeldberg, Hans Andersen Siri, and Per Nilsen Siri) arrived in Circle City on the Yukon River on 28 February 1899, with one hundred fourteen reindeer surviving out of the herd from Finnmark. After departing from the main group of the expedition at a point on the Klehini River in May 1898, they had travelled for about ten months, arduously making their way overland, driving or leading the deer (with help from Per Nilsen's dog), perhaps more than a thousand miles. Redmyer told of the journey in fine description (see Arestad, 1951).

The reindeer were used in the spring of 1899 by the United States Army in explorations in eastern and central Alaska. Shortly thereafter, when the Protestant Episcopal Mission at Golovnin Bay was planning to drive its own herd from Golovin to central Alaska, to be stationed with its department at the mouth of the Tanana River (near Weare, or Tanana), Sheldon Jackson proposed an exchange, to which the mission agreed. Coming from Circle City, the reindeer from Finnmark arrived at Tanana on 1 September 1899. The equal number of animals at Golovin, received by the Bureau of Education in exchange, was used to replace in part the herd borrowed by the federal government from Charlie Antisarlook (at Cape Prince of Wales) in the rescue of the whaling crews at Point Barrow in 1897–1898. (Jackson, 1900.)

In about 1904, the strong criticism in Alaska concerning the ownership and administration of the domesticated herds as developed by the Bureau of Education, with allegations that Sheldon

Jackson had misused funds and improperly managed the herds, was called to the attention of President Theodore Roosevelt. According to Secretary of the Interior Ethan A. Hitchcock, who stated the case in 1905, the federal government had appropriated $207,500 to import and domesticate reindeer, but in 1904 owned only 2,321 animals, while all others (5,868) were in the hands of "missionary societies, the herders, and natives" (the money mentioned did not include the special funds to relieve the Yukon gold miners). Further, that Alaskans had benefitted from the training in husbandry as it applied to their welfare was questioned. Hitchcock (authorized by the president) appointed Frank C. Churchill as special agent to investigate expenditures and modes of operation of the school service in Alaska and its program of reindeer husbandry. On that assignment, Churchill traveled in Alaska during summer 1905, gathering data; about the service and its programs, his report was quite negative, but his information could give no support to the accusations concerning misuse of federal funds (cf. Churchill, 1906). Some of the allegations may have arisen as expressions of antagonism towards the Presbyterian Church, which through its members who held determinative positions in Alaska had been fairly influential in setting educational policy and social standards in the district (cf. Nichols, 1924; Lazell, 1960). Some months after Churchill submitted his report to the Department of the Interior, Jackson resigned from the federal service, not as a consequence of the criticism of his management, but because of ill health he felt unable to continue his work. His successor, Harlan Updegraff, took office in May 1907. During following years, the program concerned with domesticated reindeer was somewhat revised, in an effort to ensure that only the indigenous Alaskans were to own such livestock.

Reindeer husbandry continued as a relatively productive industry in western Alaska, mostly carried on by companies that received deer from owners and paid out profits accordingly, yet certain problems seemed to bring about inconsistent results in management of herds. Lantis (1950) reviewed its status and, in considering some of the factors that affected positively and negatively the maintenance of the deer, observed that the Eskimos culturally were hunters on the sea rather than herders inland, and for them, a change of way of life was "difficult and disturbing." A quarter of a century later, regional corporations were organized, through which the local people as groups conceivably could have an entrepreneurial share in managing their various interests, including their reindeer enterprises; some of the corporations own fairly large herds on the mainland and smaller numbers on islands (e.g., Nunivak Island, St. Paul Island in the Pribilofs, and St. Lawrence Island). Reindeer meat is sold widely in the state; the skins no longer have much importance for clothing or household use but are valued as basic material in folk art (for example, the masks of deer skin made by the people of Anaktuvuk Pass). A small income is derived from sale of reindeer antlers, which are desirable in traditional medicine in certain areas of Asia.

Eaton Station was a management headquarters for the reindeer herds in the coastal highlands near central Norton Sound for approximately ten years. Axel E. Karlson of the Swedish Mission church and school in Unalakleet village served as superintendent (1903–1907), succeeded by William T. Lopp, who administered the entire northern region. Eventually the buildings at Eaton Station were dismantled and moved to Unalakleet, and sold to the Swedish Mission there. (Leonard Brown, personal communication; Updegraff, 1909.) Perhaps, as Carl Johan seemed to suggest in August 1898, the site of the station at eight miles from the sea was too far inland to permit efficient operation as a regional center of reindeer husbandry under the conditions prevailing in Alaska. On the other hand, Unalakleet, lying on the shore of Norton Sound, was conveniently reachable, and that village ultimately became the administrative headquarters for management of the local deer herds and the site of the Reindeer Experiment Station established in 1920 by the Bureau of Biological Survey, U.S. Department of Agriculture.

Of the staff at Eaton Station in 1898–1899, twenty-four of the people from Finnmark returned to Norway in 1900; among the others, four deaths occurred (1899–1902), including one of Carl Johan's closest friends, Johan Petter Johannesen, who on a mail run became lost on the trail in a snowstorm and died of exposure. (His team of reindeer survived.) (Jackson, 1901, 1903.)

After a century, almost no trace remains of Eaton Station. The Unalakleet River has washed away most of the site on its northern banks where once the administration building, the store, the clinic, and the barracks cabins were located, and where Carl Johan walked down to fetch water. The slope has been transformed by a new channel carved out by the river over a hundred years of frost and warming, of spring breakups of the ice, and summer and autumn floods. The vegetation has recovered; shrubby willows, and a few alders, extend along the river's edge and inland, interspersed with birch trees and white spruce on the higher ground just to north. There in the more open forest, patches of reindeer moss are growing luxuriantly, no longer grazed by the deer.

# The Journal: A Brief History

Carl Johan did not record the year when he lent his journal to a friend who asked to read his narrative of the Yukon Relief Expedition. The loan conceivably took place around 1920, but that probable date is important only as a marker of its loss to the author, for the friend, when asked, did not return the manuscript. After many requests over the passage of time that it be returned, she told Carl Johan that she had lost it. He deeply regretted its loss; he valued his journal as a historical record and with the concurrence of his spouse had thought as well of publishing the account himself.

Why the borrower did not return the manuscript is not known. Her daughter later came upon it amongst her deceased mother's possessions. She made no effort to find the author, but she translated into English portions of the work and showed her results to Keith Murray, then of the Faculty of Social Studies of Western Washington College of Education. Shortly thereafter (as he informed Carl Johan), Murray noticed the name of the author given in an article, in a newspaper from Oregon, about Leonhard Seppala, who had gained wide recognition for his part in carrying serum to Nome (during the epidemic of diphtheria there) by dog team in winter of 1925; his lead dogs Balto and Togo shared his fame. Leonhard Seppala and Carl Johan were lifelong friends, since their youth in Norway. Murray then contacted Carl Johan (1958), and eventually the manuscript was returned to its author. (An article concerning Leonhard Seppala was published in the *Oregon Journal* in 1957.)

Carl and Hannah, circa 1940.

(1958), and eventually the manuscript was returned to its author. (An article concerning Leonhard Seppala was published in the *Oregon Journal* in 1957.)

Carl Johan felt gratitude to Murray. He continued to hope that he would be able himself to publish his manuscript, and evidently he felt some apprehension in that regard. He died before he completed his editorial efforts. For his own work, Murray (1988) utilized parts of the unauthorized translation.

Karl Johan Sakariassen Fladstrand was born on 12 March 1875 at Langfjord, Norway. In America, he chose to simplify his name and the spelling of his patronym to Carl Johan Sacarisen. He was a dedicated citizen, with an open interest in the affairs of his adopted country. He and his wife, Hannah (Johannah) Lumijärvi, lived in Oregon; they were married in 1904. Over the years, Carl Johan was postmaster, farmer, and pastor of the Lutheran Church. Hannah Lumijärvi Sacarisen died on 7 January 1955, in Portland, Oregon. There, Carl Johan died on 6 August 1968.

# References Cited

Alger, R. A. Report of the Secretary of War to the Senate, 15 December 1897. 55th Congress, 2d Session. *Congressional Record*, Vol. 31, Pt. 1, page 152. Government Printing Office, Washington, D.C. 1897.

______. Letter from the Secretary of War, making report of his action under the act authorizing him to assist in the relief of people in the Yukon River country. 20 February 1899. House of Representatives. *House Documents*, Vol. 70, No. 244, 55th Congress, 3d Session. (Serial 3812.) Government Printing Office, Washington, D.C. 1899.

Andenaes, Johannes. *The general part of the criminal law of Norway*. Translated by Thomas P. Ogle. Fred B. Rothman and Company, South Hackensack, New Jersey. 1965.

Appleton, Thomas E. *Ravenscrag: The Allan Royal Mail Line*. McClelland and Stewart Ltd., Toronto. 1974.

Arestad, Sverre. Reindeer in Alaska. *Pacific Northwest Quarterly* 42: 211–223. 1951.

Bailey, Joseph W. Remarks to the House of Representatives. 16 December 1897. 55th Congress, 2d Session. *Congressional Record*, Vol. 31, Pt. 1, page 212. Government Printing Office, Washington, D.C. 1897.

Baker, W. A., and Tryckare, Tre. *The engine powered vessel*. Grosset and Dunlap, New York. 1965.

Beach, Hugh. The Saami in Alaska: Ethnic relations and reindeer herding. Pages 1–81 in *Contributions to circumpolar studies*. H. Beach, Editor. Department of Cultural Anthropology, University of Uppsala, Sweden. 1986.

Bender, Norman J. *Winning the West for Christ. Sheldon Jackson and Presbyterianism on the Rocky Mountain Frontier, 1869–1880*. University of New Mexico Press, Albuquerque. 1996.

Bjork, Kenneth O. Reindeer, gold, and scandal. *Norwegian-American Studies* 30: 130–195. 1985.

Board of Trade. Eng. 1, Agreement and account of crew, foreign-going ship: *Manitoban*, O. N. 52746. Issued by the Board of Trade, 15.1.98. Glasgow, U.K. 1898.

Bockstoce, John. The Point Barrow Refuge Station. *The American Neptune* 3: 5–21. 1979.

______. *Whales, ice, and men. The history of whaling in the western Arctic*. University of Washington Press, Seattle. 1986.

Brooks, Alfred H. *The geography and geology of Alaska*. Government Printing Office, Washington, D.C. 1906.

______. Richardson, G. B., Collier, A. J., and Mendenhall, W. C. *Reconnaissances in the Cape Nome and Norton Bay regions, Alaska, in 1900*. U.S. Geological Survey. Government Printing Office, Washington, D.C. 1901.

Brower, Charles D. *Fifty years below zero*. Dodd, Mead and Company, New York. 1960.

Bureau of Marine Inspection and Navigation. Office of U.S. Local Inspectors, Seattle. Casualties and Violations Case Files, 1887–1942. *Stranding of the Steamer "Laurada" Investigation, 9–10 March 1900*. RG 41, File No. 1086, National Archives, Pacific Alaska Region, Seattle, Washington. 1900.

Cannon, Joseph G. Address to the House of Representatives. 16 December 1897. 55th Congress, 2d Session. *Congressional Record*, Vol. 31, Pt. 1, page 211. Government Printing Office, Washington, D.C. 1897.

Carlisle, Henry. Jafet Lindeberg. An interview. *Mining Engineering* 16: 112–113. 1964.

Carlson, T. H. [*sic*] [L. H.] The discovery of gold at Nome, Alaska. *Pacific Historical Review* 15: 259–278. 1946.

______. The first mining season at Nome, Alaska—1899. *Pacific Historical Review* 16: 163–175. 1947.

______. Nome: from mining camp to civilized community. *Pacific Northwest Quarterly* 38: 233–242. 1947a.

______. *An Alaskan gold mine. The story of No. 9 Above*. Northwestern University Press, Evanston, Illinois. 1951.

*Chicago Tribune*. 1 March 1898, page l; 3 March 1898, page 4.

Churchill, Frank C. Report on the condition of educational and school service and the management of reindeer service in the District of Alaska. *Senate Documents*, Vol. 23, No. 483, 59th Congress, lst Session. (Serial 4931.) Government Printing Office, Washington, D.C. 1906.

Cocke, Mary, and Cocke, Albert. Hell roaring Mike: a fall from grace in the frozen North. *Smithsonian* 13: 119–137. 1983.

Collier, Arthur J., Hess, Frank L., Smith, Philip S., and Brooks, Alfred H. *The gold placers of parts of Seward Peninsula, Alaska, including the Nome, Council, Kougarok, Port Clarence, and Goodhope Precincts*. Department of the Interior, United States Geological Survey, Bulletin No. 328. Government Printing Office, Washington, D.C. 1908.

Cullum, George W. *Biographical register of the officers and graduates of the U.S. Military Academy*. Houghton, Mifflin and Company, Boston. 1891 et sqq.

*Dagsposten* (Trondheim). 22 January 1898.

Dawson, N. H. R. Education in Alaska. Pages 8–11 in *Report of the Commissioner of Education for the year 1887–88*. U.S. Bureau of Education, Department of the Interior. Government Printing Office, Washington, D.C. 1889.

Devore, Daniel B. Documentary File. 1885–1930. Military Reference Branch, RG 94, National Archives, Washington, D.C.

Dexter, John A. Affidavit before the United States Senate. 20 April 1900. 56th Congress, 1st Session. *Congressional Record*, Vol. 33, Pt. 5, page 4471. Government Printing Office, Washington, D.C. 1900.

Dunham, Sam C. The Alaskan gold fields and the opportunities they offer for capital and labor. *Bulletin of the Department of Labor*, No. 19. Government Printing Office, Washington, D.C. 1898.

______. The Yukon and Nome Gold Regions. *Bulletin of the Department of Labor*, No. 29. Government Printing Office, Washington, D.C. 1900.

Evensen, Arne. *Social defense in Norway*. Published by the Ministry of Social Affairs and the Ministry of Justice. Oslo. 1965.

Finnish American Society of the West. *The translated diary of Wilhelm Basi*. Special Historical Tract Issue, Vol. 6, No. 4, 28 pages. 1971.

*Finnmarksposten*. Hammerfest. 21 December 1897; 15 February 1898.

Fortuine, Robert. *Chills and fever. Health and disease in the early history of Alaska*. University of Alaska Press, Fairbanks. 1992.

*Frank Leslie's Illustrated Newspaper*. New York. 17 March 1898.

Furman, B. *A profile of the United States Public Health Service 1798–1948*. Government Printing Office, Washington, D.C. 1973.

Graarud, Aage, and Irgens, Kristen. *H. Mohn: Atlas de climat de Norvège*. Grondahl and Sons, Boktrykkeri, Kristiania. 1921.

Greeley, Adolphus W. *Handbook of Alaska*. Scribner's, New York. 1909.

Greene, Jerome A. *Yellowstone command. Colonel Nelson A. Miles and the Great Sioux War, 1876–1877*. University of Nebraska Press, Lincoln, Nebraska. 1991.

Gruening, Ernest. *The State of Alaska*. Random House, New York. 1954.

Hocking, Charles. *Dictionary of disasters at sea during the age of steam, including sailing ships and ships of war lost in action. 1824–1962*. Lloyd's Register of Shipping, Sussex, England. 1969.

Hood, Bryan C. Sacred pictures and rocks: ideological and social space in the North Norwegian Stone Age. *Norwegian Archaeological Review* 21: 65–84. 1988.

Hughes, Charles Campbell. *An Eskimo village in the modern world*. Cornell University Press, Ithaca, New York. 1960.

Hulley, Clarence C. *Alaska past and present*. Binfords and Mort, Portland, Oregon. 1958.

Hungerford, E. *Wells Fargo. Advancing the American Frontier*. Random House, New York. 1949.

Jackson, Sheldon. Letter to the Commissioner of Education. Pages xxxi–xxxii in *Report of the Commissioner of Education for the year 1877*. Government Printing Office, Washington, D.C. 1879.

______. *Alaska, and Missions on the North Pacific Coast*. Dodd, Mead and Company. New York. 1880.

______. Schools in Alaska. Pages 750–753 in *Report of the Commissioner of Education for the year 1885–86*. U.S. Bureau of Education, Department of the Interior. Government Printing Office, Washington, D.C. 1887.

______. Report to the Territorial Board of Education. Sitka, Alaska, 30 June 1888. Pages 183–194 in *Report of the Commissioner of Education for the year 1887–1888*. U.S. Bureau of Education, Department of the Interior. Government Printing Office, Washington, D.C. 1889.

______. Education in Alaska. Chapter XVII (pages 1244–1300) in *Report of the Commissioner of Education for the year 1889–90*. Vol. 2. Government Printing Office, Washington, D.C. 1893.

______. Report on education in Alaska. Chapter XXV (pages 923–960) in *Report of the Commissioner of Education for the year 1890–'91*. Vol. 2. Government Printing Office, Washington, D.C. 1894.

______. Report on educational affairs in Alaska. Chapter IX (pages 1705–1745) in *Report of the Commissioner of Education for the year 1892–93*. Vol. 1. Government Printing Office, Washington, D.C. 1895.

______. Report on education in Alaska. Chapter XXXIII (pages 1425–1455) in *Report of the Commissioner of Education for the year 1894–95*. Vol. 2. U.S. Bureau of Education. Government Printing Office, Washington, D.C. 1896.

______. *Report on introduction of domestic reindeer into Alaska, with illustrations. 1896.* (*Senate Documents,* Vol. 3, No. 49, 54th Congress, 2d Session.) (Serial 3469.) Government Printing Office, Washington, D.C. 1897.

______. Memorandum to the House of Representatives, 16 December 1897. 55th Congress, 2d Session. *Congressional Record*, Vol. 31, Pt. 1, pages 210–211. Government Printing Office, Washington, D.C. 1897a.

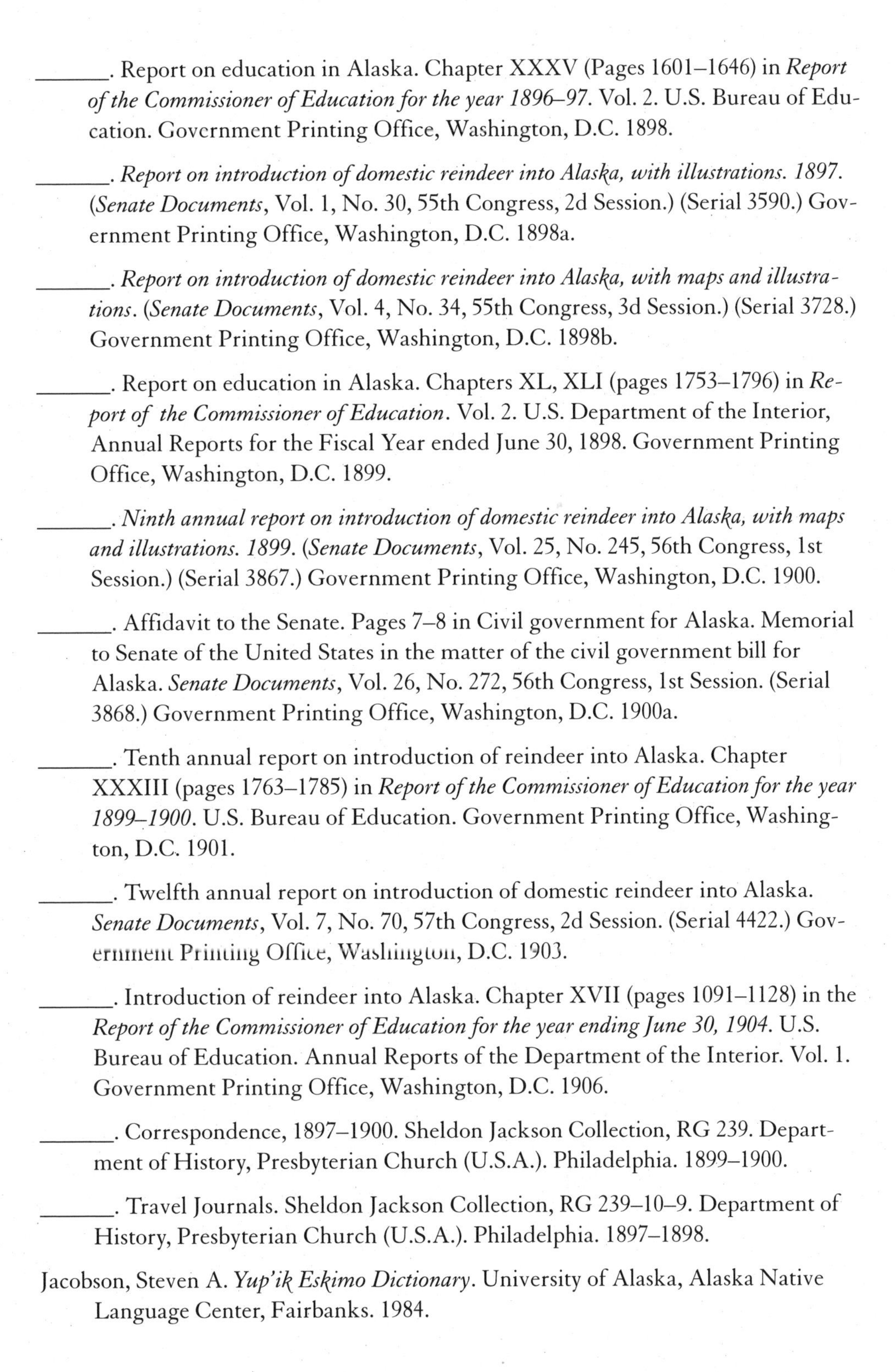

______. Report on education in Alaska. Chapter XXXV (Pages 1601–1646) in *Report of the Commissioner of Education for the year 1896–97*. Vol. 2. U.S. Bureau of Education. Government Printing Office, Washington, D.C. 1898.

______. *Report on introduction of domestic reindeer into Alaska, with illustrations. 1897.* (*Senate Documents*, Vol. 1, No. 30, 55th Congress, 2d Session.) (Serial 3590.) Government Printing Office, Washington, D.C. 1898a.

______. *Report on introduction of domestic reindeer into Alaska, with maps and illustrations*. (*Senate Documents*, Vol. 4, No. 34, 55th Congress, 3d Session.) (Serial 3728.) Government Printing Office, Washington, D.C. 1898b.

______. Report on education in Alaska. Chapters XL, XLI (pages 1753–1796) in *Report of the Commissioner of Education*. Vol. 2. U.S. Department of the Interior, Annual Reports for the Fiscal Year ended June 30, 1898. Government Printing Office, Washington, D.C. 1899.

______. *Ninth annual report on introduction of domestic reindeer into Alaska, with maps and illustrations. 1899.* (*Senate Documents*, Vol. 25, No. 245, 56th Congress, 1st Session.) (Serial 3867.) Government Printing Office, Washington, D.C. 1900.

______. Affidavit to the Senate. Pages 7–8 in Civil government for Alaska. Memorial to Senate of the United States in the matter of the civil government bill for Alaska. *Senate Documents*, Vol. 26, No. 272, 56th Congress, 1st Session. (Serial 3868.) Government Printing Office, Washington, D.C. 1900a.

______. Tenth annual report on introduction of reindeer into Alaska. Chapter XXXIII (pages 1763–1785) in *Report of the Commissioner of Education for the year 1899–1900*. U.S. Bureau of Education. Government Printing Office, Washington, D.C. 1901.

______. Twelfth annual report on introduction of domestic reindeer into Alaska. *Senate Documents*, Vol. 7, No. 70, 57th Congress, 2d Session. (Serial 4422.) Government Printing Office, Washington, D.C. 1903.

______. Introduction of reindeer into Alaska. Chapter XVII (pages 1091–1128) in the *Report of the Commissioner of Education for the year ending June 30, 1904*. U.S. Bureau of Education. Annual Reports of the Department of the Interior. Vol. 1. Government Printing Office, Washington, D.C. 1906.

______. Correspondence, 1897–1900. Sheldon Jackson Collection, RG 239. Department of History, Presbyterian Church (U.S.A.). Philadelphia. 1899–1900.

______. Travel Journals. Sheldon Jackson Collection, RG 239–10–9. Department of History, Presbyterian Church (U.S.A.). Philadelphia. 1897–1898.

Jacobson, Steven A. *Yup'ik Eskimo Dictionary*. University of Alaska, Alaska Native Language Center, Fairbanks. 1984.

Kittilsen, Albert N. Affidavit to the Senate. Pages 8–9 in Civil government for Alaska. Memorial to Senate of the United States in the matter of the civil government for Alaska. *Senate Documents*, Vol. 26, No. 272, 56th Congress, 1st Session. (Serial 3868.) Government Printing Office, Washington, D.C. 1900.

Kratzer, Melissa. Liberia illustrated. American pioneers abroad. *American History Illustrated* 16: 46–49. 1981.

Lagercrantz, Eliel. *Lappischer Wörterschatz*. Vols. 1, 2. Druckerei-A.G. der Finnischen Literaturgesellschaft, Helsinki. 1939.

Lane, Charles D. Brief on behalf of Charles Lane and other American citizens, owners of mining claims in Alaska. Pages 1–7 in Civil government for Alaska. Memorial to Senate of the United States in the matter of the civil government bill for Alaska. *Senate Documents*, Vol. 26, No. 272, 56th Congress, 1st Session. (Serial 3868.) Government Printing Office, Washington, D.C. 1900.

Lantis, Margaret. The reindeer industry in Alaska. *Arctic* 3: 27–44. 1950.

Lazell, J. Arthur. *Alaskan Apostle*. Harper and Brothers, New York. 1960.

Levina, M. G., and Potapova, L. P. *Istoriko-etnograficheskii Atlas Sibiri*. Akademiia Nauk SSSR, Moskva. 1961.

Lindeberg, Jafet. Affidavit to the Senate. Page 8 in Civil government for Alaska. Memorial to Senate of the United States in the matter of the civil government bill for Alaska. *Senate Documents*, Vol. 26, No. 272, 56th Congress, 1st Session. (Serial 3868.) Government Printing Office, Washington, D.C. 1900.

Lindley, Curtis H. *A treatise on the American law relating to mines and mineral lands within the public land states and territories and governing the acquisition and enjoyment of mining rights in lands of the public domain*. Vol. 1. Bancroft-Whitney Company, San Francisco. 1897.

*Lloyd's Weekly Shipping Index*. Corporation of Lloyd's, London. 1898.

*Maritime Register of New York*. 5 January–10 August 1898. Vol. 30. 1898.

Matthews, Frederick C. *American merchant ships 1850–1900*. Marine Research Society, Salem, Massachusetts. 1930.

Merriam, Henry C. Report of Maj. Gen. Henry C. Merriam, Commanding the Department of the Columbia. Pages 179–181 in Annual reports of the War Department for the Fiscal Year ended June 30, 1898. Vol. 1, Pt. 2. Report of the Major General Commanding the Army. *House Documents*, Vol. 3, No. 2, 55th Congress, 3d Session. (Serial 3745.) Government Printing Office, Washington, D.C. 1898.

Michael, Henry N. (Ed.) *Lieutenant Zagoskin's travels in Russian America, 1842–1844*. University of Toronto Press, Toronto. 1967.

Muir, John. *The Cruise of the Corwin. Journal of the Arctic Expedition of 1881 in search of DeLong and the Jeannette.* Houghton Mifflin Company, Boston and New York. 1917.

Murray, Keith A. *Reindeer and gold.* Occasional Paper No. 24, Center for Pacific Northwest Studies, Western Washington University, Bellingham, Washington. 1988.

Nelson, E. W. *The Eskimo about Bering Strait.* 18th annual report of the Bureau of American Ethnology, 1896–1897, Part 1. Government Printing Office, Washington, D.C. 1899.

Newell, Gordon. (Ed.) *The H. W. McCurdy Marine History of the Pacific Northwest.* Superior Publishing Company, Seattle. 1966.

Newell, Gordon, and Williamson, J. *Pacific coastal liners.* Superior Publishing Company, Seattle. 1959.

*New York Maritime Register.* 18 October 1899.

*New York Sun.* 28 February 1898.

*New York Times.* 9 November 1897, page 5; 1 December 1897, p. 1; 28 February 1898, p. 5.

Nichols, Jeannette Paddock. *Alaska. A history of its administration, exploitation, and industrial development during its first half century under the rule of the United States.* Arthur H. Clark Company, Cleveland. 1924.

Nikolaevskii, L. D. Diseases of reindeer. Chapter 8 (pages 230–293) in *Reindeer husbandry* [*Severnoe olenevodstvo*]. P. S. Zhivunov, Editor. (Translated from Russian.) Israel Program for Scientific Translations Ltd., Jerusalem. 1968.

*Nome Gold Digger.* 25 October 1899; 1 November 1899; 4 November 1899.

Norris, Frank B. *Legacy of the gold rush: An administrative history of Klondike Gold Rush National Historical Park.* National Park Service, Alaska Systems Support Office, Anchorage, Alaska. 1996.

*Oregonian.* Portland, Oregon. [*The Oregonian.*] 28 November 1897.

*Oregon Journal.* Section 4, page 1. Portland, Oregon. 19 September 1957.

Orth, Donald J. *Dictionary of Alaska Place Names.* Geological Survey Professional Paper 567. Government Printing Office, Washington, D.C. 1971.

Qvigstad, J. *Den tamme rens sykdommer.* K. Karlsens Trykkeri. Tromsø. 1941.

Rausch, R. L. Tropical problems in the Arctic: Infectious and parasitic diseases, a common denominator. Pages 63–70 in *Industry and Tropical Health: VIII.* Ruth Pelizzon, Editor. Harvard School of Public Health, Boston. 1974.

Ray, Dorothy Jean. *The Eskimo of St. Michael and vicinity as related by H. M. W. Edmonds*. Anthropological Papers, University of Alaska Fairbanks, Department of Anthropology, College, Alaska. 1966.

Ray, P. H. Letter to the Adjutant-General, U.S.A. 6 September 1897. Page 13 in Alger, R. A. Letter from the Secretary of War, making report of his action under the act authorizing him to assist in the relief of people in the Yukon River country. House of Representatives, 20 February 1899. *House Documents*, Vol. 20, No. 244, 55th Congress, 3d Session. (Serial 3812.) Government Printing Office, Washington, D.C. 1899.

______. Letter to the Adjutant-General, United States Army. 18 December 1897. Pages 94–95 in Alger, R. A. Letter from the Secretary of War, making report of his action under the act authorizing him to assist in the relief of people in the Yukon River country. House of Representatives, 20 February 1899. *House Documents*, Vol. 20, No. 244, 55th Congress, 3d Session. (Serial 3812.) Government Printing Office, Washington, D.C. 1899a.

Rayburn, Alan. *Naming Canada*. University of Toronto Press, Toronto. 1994.

*San Francisco Chronicle*. 27 October 1897, page 1; 29 October 1897, page 2; 4 November 1897, page 5, et seq.

Schjoldager, Harald, and Backer, Finn. *The Norwegian Penal Code. With an introduction by Professor jur. Johs. Andenaes*. (Translated to English by the authors.) Fred B. Rothman and Company, South Hackensack, New Jersey. 1961.

Schrader, Frank C., and Brooks, Alfred H. *Preliminary report on the Cape Nome gold region, Alaska*. Department of the Interior, U.S. Geological Survey. Government Printing Office, Washington, D.C. 1900.

*Seattle Daily Times*. 9 March 1898, page 1; 15 March 1898, page 9; 19 October 1898.

*Seattle Post-Intelligencer*. 28 October 1897, page 4; 7 February 1898; 12 March 1898, page 8; 15 March 1898; 7 April 1898, page 6; 9 April 1898; 16 April 1898; 29 May 1898, page 16; 22 June 1898, page 3; 13 July 1898, page 9; 29 June 1898, page 12; 16 October 1899, p. 1.

Shapiro, Mary. *A picture history of the Brooklyn Bridge*. Dover Publications, New York. 1983.

Stürchler, D. *Endemic areas of tropical infections*. Éditiones Roche, Hans Huber Publishers, Bern. 1988.

Thuland, C. M. Letter, 15 June 1902. Pages 158–160 in Clausen, C. A. Life in the Klondike and Alaska gold fields. *Norwegian-American Studies and Records* Vol. 16. 1950.

*Trondhjem's Adresseavis*. 24 January 1898; 5 February 1898.

Tuttle, F. Report of F. Tuttle, Captain, R. C. S., Commanding U.S. Revenue Cutter *Bear*. Pages 128–137 in *Report of the cruise of the U.S. Revenue Cutter* Bear *and the overland expedition for the relief of whalers in the Arctic Ocean, from November 27, 1897, to September 13, 1898.* Division of the Revenue Cutter Service. Treasury Department Document 2101. (*House Documents*, Vol. 93, No. 511, 56th Congress, 2d Session.) (Serial 4167.) Government Printing Office, Washington, D.C. 1899.

United States Congress. House of Representatives. Annual report of the Governor of Alaska. [House ExDoc 1, through 1894.] *House Executive Documents*, Vol. 12, No. 1, 48th Congress, 2d Session. (Serial 2287.) Government Printing Office, Washington, D.C. 1884.

______. House of Representatives. Annual report of the Governor of Alaska. [House Doc 5, 1895 et sqq.] *House Documents*, Vol. 16, No. 5, 54th Congress, 1st Session. (Serial 3383.) Government Printing Office, Washington, D.C. 1895.

______. House of Representatives. Discussion, 15 January 1889. 50th Congress, 2d Session. *Congressional Record*, Vol. 20, page 798. Government Printing Office, Washington, D.C. 1889.

______. House of Representatives. Debates, 16 December 1897. 55th Congress, 2d Session. *Congressional Record,* Vol. 31, Pt. 1, pages 210–213. Government Printing Office, Washington, D.C. 1897.

______. Senate. Debates, 9 December 1897. 55th Congress, 2d Session. *Congressional Record*, Vol. 31, Pt. 1, page 47. Government Printing Office, Washington, D.C. 1897a.

______. Senate. Hearings, 7–28 April 1900. 56th Congress, 1st Session. *Congressional Record*, Vol. 33, Pt. 5. Government Printing Office, Washington, D.C. 1900.

______. Senate. Military Affairs Committee. (Thomas H. Carter, Chairman.) *Compilation of narratives of explorations in Alaska*. (*Senate Reports,* Vol. 11, No. 1023, 56th Congress, 1st Session.) (Serial 3896.) Government Printing Office, Washington, D.C. 1900.

United States Department of Transportation. Office of the Assistant Secretary for Public Affairs: U.S. Coast Guard, *News* 34–97, 11–13–97. Washington, D.C. 1997.

United States Department of the Treasury. *Warrants, 1801–1921.* Warrant No. 1913. RG 11, General Records of the U.S. Government. 1897.

______. *Report of the cruise of the U.S. Revenue Cutter* Bear *and the overland expedition for the relief of the whalers in the Arctic Ocean from November 27, 1897, to September 13, 1898.* Division of the Revenue Cutter Service, Treasury Department Document No. 2101. (*House Documents,* Vol. 93, No. 511, 56th Congress, 2d Session.) (Serial 4167.) Government Printing Office, Washington, D.C. 1899.

______. Letter, 11 May 1899, from the Acting General Superintendent, Life-Saving Service. RG 36, Requests for Wreck-Reports, 1876–1903. National Archives Pacific-Alaska Region, Seattle, Washington. 1899a.

United States Department of War. Annual reports of the War Department. *House Documents*, Vol. 3, No. 2, 55th Congress, 3d Session. (Serial 3745.) Government Printing Office, Washington, D.C. 1898.

______. Annual reports of the War Department for the Fiscal Year ended June 30, 1899. Report of the Major-General Commanding the Army. Vol. 1, Pt. 3. Report of Maj. Gen. William R. Shafter, Commanding the Department of the Columbia. *House Documents*, Vol. 4, No. 2, 56th Congress, 1st Session. (Serial 3901.) Government Printing Office, Washington, D.C. 1899.

______. Office of the Adjutant General, AGO File 67221. RG 94, National Archives, Washington, D.C. 1897, 1898, 1899.

United States Post-Office Department. Report of the Postmaster-General. *Annual reports of the Post-Office Department*. Government Printing Office, Washington, D.C. 1898.

______. Report of the Postmaster-General. *Annual reports of the Post-Office Department*. Government Printing Office, Washington, D.C. 1899.

______. Report of the Postmaster-General. *Annual reports of the Post-Office Department*. Government Printing Office, Washington, D.C. 1900.

*United States Reports*. Reilly and another *v*. Campbell and others. Vol. 116, pages 418–423. Banks and Brothers, Law Publishers. New York. 1886.

Updegraff, Harlan. Report on the Alaska School Service and on the Alaska Reindeer Service. Chapter XV, pages 371–411 in *Report of the Commissioner of Education for the year ended June 30, 1907*. Vol. 1. Government Printing Office, Washington, D.C. 1908.

______. The Alaska Reindeer Service. Pages 1046–1056 in *Report of the Commissioner of Education for the year ended June 30, 1908*. Vol. 2. Government Printing Office, Washington, D.C. 1909.

*Victoria Daily Colonist*. (British Columbia.) 27 October 1897, page 5; 4, 23 April 1898.

Vorren, Ørnulv. *Saami, reindeer, and gold in Alaska. The emigration of Saami from Norway to Alaska*. Waveland Press, Prospect Heights, Illinois. 1994.

Walker, E. S. Report to the Adjutant-General, United States Army. 10 October 1898. Pages 82–84 in Annual reports of the War Department for the Fiscal Year ended June 30, 1899. Report of the Major-General Commanding the Army. Vol. 1, Pt. 3. Report of Maj. Gen. William R. Shafter, Commanding the Department of the Columbia. *House Documents*, Vol. 4, No. 2, 56th Congress, 1st Session. (Serial 3901.) Government Printing Office, Washington, D.C. 1899.

Wells, H. Hazard. Report. Pages 114–123 in Alger, R. A. Letter from the Secretary of War, making report of his action under the act authorizing him to assist in the relief of people in the Yukon River country. House of Representatives, 20 February 1899. *House Documents*, Vol. 70, No. 244, 55th Congress, 3d Session. (Serial 3812.) Government Printing Office, Washington, D.C. 1899.

______. *Magnificence and misery, a firsthand account of the 1897 Klondike Gold Rush*. Randall M. Dodd, Editor. Doubleday and Company, Garden City, New York. 1984.

Wickersham, James. *Old Yukon. Tales—trails—and trials*. Washington Law Book Company, Washington, D.C. 1938.

Williams, Gerald O. *Michael J. Healy and the Alaska maritime frontier*. [Michael A. Healy.] Thesis, University of Oregon. Published by University Microfilms International, Ann Arbor, Michigan. 1987.

Wines, Arthur Frederick. Report on the Cape Nome Mining region. *Senate Documents*, Vol. 33, No. 357, 56th Congress, 1st Session. (Serial 3875.) Government Printing Office, Washington, D.C. 1900.

Wirt, Loyal Lincoln. *Alaskan adventures*. Fleming H. Revell Company, New York. 1937.

*World* (New York). (Newspaper.) 28 February 1898, page 9.

# Index

## A


Abercrombie, W. R. (capt., 2d Infantry), 214

Abrahamsen, Jeremias, 74, 84, 93, 105, 106, 118, 120, 121, 198–200

Adams, Fort, 11

Adams, Gabriel, 225

Agricultural Acts, U.S. Congress, 10, 49

Agriculture, British Dept. of, 24, 25

Agriculture, Dept. of, secretary of, 11, 212, 235

Akutan Pass, 125, 208

Alaska Commercial Co., 8, 50, 52, 53, 126, 195, 197, 207, 220, 221, 225

Alaska, District of, 49, 220, 234

Alaska Exploration Co., 197, 230

Alaska, Organic Act, 6, 224

*Alaska* (whaling ship), 229

Alger, Russell A. (secretary of war), 15, 16, 19–21, 23, 24, 27–29, 42, 44, 45, 47, 48, 51–55, 57, 58, 212, 213

Alice (*see* Omekejook), 49, 166, 219, 223

Alien Act of 1887, 228

Aliens and noncitizens: mining claims, 192, 197, 201, 202, 225, 228, 229

Allan Steamship Line (Allan Line), 28–30, 57, 211, 213

Allen, C. E. (capt., *Del Norte*), 1, 218

*Aloha* (ship), 229

Alta, Finnmark Province, 4, 19, 20, 26–28, 30, 32–34, 38, 40, 57, 59, 63, 64, 131, 211

American Missionary Society, 198

American Prayer Day (Thanksgiving Day), 155–159, 222

Anaktuvuk Pass, artwork, 235

Andersen, Sofie (from Karasjok), 36

Anderson, August (teacher, Unalakleet), 219

Anderson, P. H. (missionary), 156, 158, 168, 186, 191, 219, 223, 225

Anderson, Thomas M. (col., CO, 14th Infantry), 215, 217

Andreafski (village), 11

Anglican Church of Canada, 8

Antisarlook, Charlie (owner of reindeer), 14, 51, 233

Anvik, 170

Anvil City (Nome), 179, 188, 194, 195, 197, 199, 206, 223, 226–230

Anvil Creek, 156–158, 190–193, 196, 198, 199, 201, 202, 205, 223–225

Anvil Glacier, 226

*Anvil News* (newspaper), 206

Arctic Ocean, 12

Arestad, Sverre, 216

Army, U.S. (*see also* U.S. Dept. of War), 12, 51, 233

Arthur, President Chester A., 51

Aubert, V. M. (consular attorney), 41–44, 61


## B


Bacon, Augustus O. (senator, Georgia), 18

Baer, Anders Aslaksen, 233

Baer, Ole (Olesen), 169, 192, 226

Bailey, Joseph W. (representative, Texas), 17–19

Bals, Inge, 111, 184

Bals, Nils, 177

Balto, Samuel, 81, 105, 106

Balto (L. Seppala's lead dog), 236

Barrow (Point Barrow), 9, 12–14, 51, 221, 233

Basi, Wilhelm, 65, 74, 80, 81, 84, 104, 142, 158, 168, 176, 188, 192, 205, 207–210, 217, 227, 231

Bautajok (town), 34

Beach, Hugh, 36

*Bear*, U.S. Revenue Marine Cutter, 8–11, 13, 14, 198, 199, 206, 221, 227

Berg, Ole, 111, 179, 188

Berg, Peder, 105, 106, 118, 148, 168, 216

Bering Sea coast: gold-bearing sand, Nome, 200, 227, 228

Berries (fruit), 128, 133, 219

*Bertha* (steamship), 205, 207, 229

Bertholf, Ellsworth P. (U.S. Revenue Marine Service), 14, 51

Biti, Anders, 115, 116

Biti, Marit, 36

Blaine, James G. (secretary of state), 11

Blake, H. L. (miner, Council City), 219

Bliss, C. N. (secretary of the interior), 20, 55

Boine, Klemet Persen, 233

Boland, Leif, 5

Bonanza District (Council City mines), 188, 224

Bonney and Stewart Chapel (Seattle), 214

Bordewich, Henry (American consul, Christiania), 25, 40–46, 213

Bossekop (harbor, Finnmark), 4, 27, 30, 32–34, 36, 40, 56, 63, 65, 66, 70, 210

Brady, John Green (governor of Alaska), 202, 229

Braes, Andrew G. (master, *Manitoban*), 28, 210

Brainard, David L. (capt., U.S. Army Qtrmaster Dept.), 215

Brevig, T. L., and Mrs. Brevig, 199

*Bristol* (steamship), 13

British Colonial Steamship Co., 57

Brooks, Alfred H. (U.S. Geol. Survey), 225, 227

Brower, Charles D. (Charlie), 51

Brown, Leonard (Unalakleet), 50

Brynteson, John, 156, 157, 159, 166, 184–186, 191–193, 196, 198, 202, 223, 224

Buchanan (lst mate, *Manitoban*), 210

Bureau of Biological Survey (U.S. Dept. Agriculture), 235

## C

Call, Samuel J. (surgeon, U.S. Revenue Marine Service), 14, 51

*Campania* (steamship), 44, 47

Campbell, E. O., and Mrs. Campbell, 199

Cannon, Joseph G. (representative, Illinois), 16, 17, 53

Cape Blossom (Mission Station, Friends), 221, 227

Cape Nome Mining Company, 197

Cape Nome Mining District, 158, 193, 194, 202, 225–227

Cape Prince of Wales, 11, 14, 135, 140, 233

Cape Smythe (Barrow), 51

Carlson, L. H., 222, 224

Carter, Thomas H. (senator, Montana), 16, 228

Cattle, disease in, 24, 25, 28, 56

Chambers, J. J. (physician, Dawson City), 222

Chandler, William E. (senator, New Hampshire), 16

Cheever, A. R. (Teller Reindeer Station), 199

Chenik (*see* Golovin)

Chicago, 77, 210

Chicago, Milwaukee, and St. Paul Railroad, 212

*Chicago Tribune* (newspaper), 212, 213

Chilkat Pass (Trail), 94, 214, 215, 218

Chilkat (Chilkot) River, 90, 91, 216

Chilkat (Chilkot) village, 90, 97, 220

Cho-kok (Chiu-kak) village, 186, 224

Christiania, 25, 38, 40–46, 213

Christmas celebration (Eaton Station), 163, 165, 166

Chukchi reindeer herders, 10, 50, 220

Churchill, Frank C., 234

Circle City, 15, 52, 83, 104, 119, 214, 216, 220, 233

*City of Seattle* (steamship), 217

Clara Creek (Seward Peninsula), 190, 201

*Cleveland* (steamship), 205, 229

Cleveland Brothers' Company, 197, 207

Cleveland (physician), 216

Clum, John P. (postal inspector), 220

Coast Seaman's Union, 50

Cockrell, Francis M. (senator, Missouri), 16

Cod fishing, winter, Unalakleet area, 222

Columbia, Dept. of, U.S. Army, 52, 213, 215, 217

Commissary General, U.S. Army, 211

Congregational Church, 14

Congress, U.S., 2, 10, 12, 14–20, 37, 47, 51–54, 127, 227–229

Constantin (*see* Uparazuck), 157, 192, 223, 225

Consul, American, in Norway, 25, 40–46, 213

Contract of Service (for Norwegian people), 35

Cooper's Station (arctic coast), 221

Copper River area, exploration, 214

Corbin, H. C. (adjutant general, Dept. of War), 42, 54

*Corwin* (U.S. Revenue Marine Cutter), 230

Council City, 139, 145, 156, 157, 162, 219, 224, 226

Court, Municipal, in Christiania (and records), 38, 41–47, 60, 61

Craigie, Wallace C. (2d lieutenant, 2d Infantry), 226

## D

*Dagsposten* (newspaper, Trondheim), 37

Dahl, Mrs., 213

Dahl, Regnor, 109, 111–117, 119, 129, 130, 132, 134, 135, 139, 142, 149, 151, 160, 161, 177, 188–190, 201, 205, 213, 218, 223, 224

Dalton, Jack (Dalton Trail, Chilkat Pass), 214

Darling Creek (Seward Peninsula), 189

Dawson City (Dawson), 11–13, 15, 25, 51, 52, 86, 107, 139, 180, 195, 197, 205, 214, 216, 222

Dawson, N. H. R. (commissioner of education), 50

Deering (gold miner)192, 198

*Del Norte* (steamship), 1–3, 121, 127–132, 135, 137, 150, 151, 179, 218, 219

Department of the Columbia, U.S. Army, 52, 213, 215, 217

Devil's club *(Oplopanax horridum)*, 216

Devore, Daniel B., 19–26, 28–33, 36–38, 40–48, 55, 57, 62, 210–214

Dexter Creek (Seward Peninsula), 192, 193, 201, 205

Dexter, John A. (merchant at Golovin), 173, 186, 224, 226, 227

Diphtheria, Nome, 236

Disease, in cattle, 24, 25, 28, 56

Disease, in reindeer, 25, 56, 167, 214, 223

Dodge, Theodore Ayrault (col., U.S. Army, Ret.) and Mrs. Dodge, 22, 55

Donnak, Tanejja (gold claim established), 136, 180, 181

Doty, William Furman (pastor), 198, 199, 219

Douglas Island, 107

Dutch Harbor, 125, 208, 220, 230

Dyea, 18, 83, 85, 86, 91, 107, 214, 215, 217, 220

## E

Easter, observance, 93, 94, 180

Eaton, John (commissioner of education), 6

Eaton Station, 6, 49, 50, 134, 135, 137, 138, 141–143, 145–147, 150–154, 156, 159–166, 168, 169, 173, 174, 176, 179–181, 186, 187, 189, 192, 199, 220–223, 235, 237, 238

Education, Alaska, 2, 6–8, 49, 50, 217

Education, Bureau of (Dept. of the Interior), 2, 11, 12, 14, 54, 219, 220, 222, 224, 233

Education, general agent of (Alaska), 1, 2, 7, 8, 20, 50

Education, U.S. Commissioner of, 6, 7, 10, 11, 50

Egavik (village), 182

Eira, Berit Nilsdatter, 36

Eira, Mathis, 34, 36, 59

El Dorado Mining District (Seward Peninsula), 139, 226

Eldridge, Bogardus (capt., 14th Infantry), 214–217

Emanuel, Round, and Nathan (London solicitors), 41, 42, 61

Engelstadt, Edwin, 133, 150, 161, 166, 220, 221

Eskimos, 64, 126, 127, 129, 131, 133, 134, 138, 150–154, 157, 161, 162, 165, 166, 174, 176, 177, 180–188, 192, 198, 201, 211, 222, 227, 234

Everett (town, Washington state), 213

## F

*Farallon* (steamship), 218

Feddersen and Nessen Agency, Hammerfest, 60

Finland, 31, 34

Finnmark, 4, 19, 20, 23, 26–28, 30, 31, 36, 37, 40, 46, 47, 76, 210, 213, 233, 235

*Finnmarksposten* (newspaper, Hammerfest), 36, 37

Fishhook, Tlingit people, 87, 88

Fishing, Unalakleet River (silver salmon), 135, 136, 138, 220

Fladstrand, 4, 5, 36, 63, 237

Fold, Julius (miner), 198

Foot-and-mouth disease, 24, 25, 56

Fort Davis (Anvil City), 226

Fort Townsend (barracks, Port Townsend), 213, 215

Fort Yukon, 15, 52

Fourth of July, U.S.A., observance, 122, 197

Friends (Quakers), Mission Station (Cape Blossom), 221

Fur seals, 50

G

Gage, Lyman J. (secretary of the treas.), 13, 51

Gambell, Francis H. (physician), 129, 130, 140, 151, 152, 161–163, 166, 168–170, 172, 173, 176, 177, 187, 199, 222

Gambell, Margaret, 219

Gambell, Nellie F., 13, 219

Gambell, Vene C., 13, 51, 219, 222

*Garonne* (steamship), 196

Gaup, Aslak Aslaksen, 136

*Geiser* (steamship), 73, 212

*General Meigs* (steamship, U.S. Army), 211

Gjesteretssag (Court, Christiania), 41

Glenn, E. F. (capt., 25th Infantry), 214

Golovin (Chenik) (village), 1, 11, 131, 157, 158, 173, 186, 194, 219, 222–225, 233

Golovnin Bay, 3, 121, 131, 137, 139, 141, 142, 145–147, 150, 151, 155–157, 162, 169, 170, 172, 174, 185, 186, 219, 222, 233

Good Hope Mining Company, 222

Great Northern Railroad, 212, 213

Greeley, Adolphus W. (maj. gen.; Army Signal Corps), 226

Greenock (port, Scotland), 28

Gregory, J. R. (physician), 192, 223 (and Mrs. Gregory), 227

Greiner, Otto, 97, 104, 135, 179

Gumaer, Alfred W. (special agent and scout, War Dept.), 211, 212, 214

Gypsy (dog), 189

H

Haetta, Ida Johannesdatter, 36

Haetta, Isak Johannesen, 36, 184

Haetta, Jakob, 111, 135

Haetta, Per Johannessen, 231, 233

Hagelin, John L., 154, 156, 158, 166, 184–186, 191, 222

Haines (Haines Mission), 86–88, 94, 107, 108, 115, 119, 122, 134, 138, 180, 214–218, 220

Hammerfest, 4, 26, 27, 32, 34, 37

Hansbrough, Henry Clay (senator, North Dakota), 228, 229 (Amendment)

Hansen, Amund, 132, 179, 188, 189, 204

Hansen, Hjalmar (Hilmar), 74, 135, 181

Harris, William T. (commissioner of education), 10, 11

Harsimus Cove (Jersey City), 212

Haster (gold miner), 171

Hatch, E. T. (deputy collector, Customs Office, St. Michael) and Mrs. Hatch, 166, 167, 223

Havana (Cuba), 217

Hay, John (ambassador to Britain; secretary of state), 24, 25, 45

Healy, John (No. Amer. Transp. and Trading Co.), 52

Healy, Mary, 9, 50

Healy, Michael A. (capt., U.S. Revenue Marine Service), 8–11, 50

Healy, Michael Morris, 50

*Healy* (icebreaker, U.S. Coast Guard), 50

Heftye, Thomas Joh. and Son (bank, Christiania), 43, 44

Hegner, Edward, 54

Hendriksdatter, Brita, 4

Hendrikson, K. J. (Swedish Evangelical Mission Covenant Church), 225

Hermansen, Alfred, 162, 167

Herschel Island, 13

Hess, Frank L. (U.S. Geol. Survey), 222

Hitchcock, Ethan A. (secretary of the interior), 234

Hobart, Garret A. (vice pres., U.S.), 19

*Homer* (ship), 229

Hospice of St. Bernard (Nome), 229

House Resolution 5173 (U.S. Congress), 16–19, 53, 54

Hovick, A. (Teller Reindeer Station), 199

Huldtberg, Nels O. (missionary, Swedish Evangelical Mission Covenant Church, and gold miner), 156–158, 191, 219, 222, 223, 225

*Humboldt* (steam schooner), 218

Hummel, John (gold miner, Nome beach sands), 227

Hunter, John (baker, on steamship *Manitoban*), 210

Hvidsted, Jens C. (also Widstead, Widsted) (N. Amer. Transp. & Trading Co.), 170, 192, 224

## I

Iceland, 67

Ilwaco (town, Washington State), 209

Inari (town, Scandinavia), 34

Interior, Dept. of, 2, 6, 12, 20, 54, 199, 217, 219, 222, 224, 233, 234

Interior, secretary of the, 6, 12, 20, 49, 55, 234

Iñupiaq, names (spelling), 224

Ireland, 44

Ivanof, Stefan (gold claim, Anvil Creek), 192

## J

Jackson, Sheldon, 1–5, 7–13, 16–27, 29, 30, 32–38, 40, 41, 43, 45–51, 53, 55, 56, 58, 59, 64, 68, 70, 71, 85, 89, 113, 114, 116–121, 126, 128–132, 135–137, 146, 149, 156, 198–200, 210–222, 227–229, 233, 234

*Jane Gray* (clipper whaler), 219, 222

Jansen, Max (Janson), 24, 25, 29, 30, 32, 33, 40–46, 58–62, 212

Jarvis, David H. (lt., U.S. Revenue Marine Service), 14, 51, 227

Joernes, Alex, 134, 152, 161, 189,190

*J. J. Healy* (river steamer), 12

Johannesen, Johan Petter (Peder), 132, 177, 179,181, 187, 189, 190, 192, 224, 235

Johannessen, Sakarias, 4

Johnson, Charles S. (judge), 229

Johnson, David (teacher, Unalakleet), 219

Johnson, Malvina (teacher, Unalakleet), 49, 154, 165, 166, 219, 223

Josefsen, Samuel (Lakselv), 104, 105

Juneau (town), 107, 122, 123, 218

## K

Karasjok, 34, 36

Karlson, Axel E. (missionary, Swedish Evangelical Mission Covenant), 49, 133, 158, 162, 165, 166, 168, 173–176, 181, 191, 202, 219, 224, 235

Karlson, Hanna Swenson, 154, 219

*Karluk* (steam brigantine), 13

Kautokeino, 27, 31, 34, 36, 213, 219

Kelly, John W. (purchasing agent), 220

Kelly, Luther S. (capt.; special agent and scout, Dept. of War), 211, 212, 214

Kemi, Samuel Johnsen, 34, 36, 59

Keskitalo, Jan Henry, 36

Kierulff, H. Newton (physician, Anvil City), 229

Kimball, A. S. (lt. col., Depot Quartermaster, New York), 211

Kimball Agency, J. S., 220

King, F. (physician), 198, 203, 204

Kittilsen, Albert N. (physician and miner), 130, 156, 157, 159, 160, 169, 170, 172, 184, 191, 193, 209, 219, 222, 223, 228

Kjeldberg, Emil, 233

Kjeldsberg, Magnus, 59, 109, 114, 116, 128, 135, 142, 158, 161, 170, 171, 188, 192, 195, 207, 209, 227

Kjeldsberg, Thoralf, 75, 118, 128, 135, 163, 165, 176, 188, 192, 207, 209, 227

Kjellmann, William A. (Wilhelm), 3, 4, 6, 11, 12, 15, 19, 20, 22, 23, 25–27, 29–34, 36–38, 40, 43, 46, 47, 49, 50, 54, 56, 59, 60, 64, 65, 68–72, 74, 80–82, 84–94, 96–107, 109, 119–121, 129–132, 134–139, 141, 143–152, 155, 158–161, 163, 166–168, 170–173, 177, 179–181, 188, 190, 192, 195, 199, 205, 209, 210, 213–216, 219–224, 227

Kjellmann, Torvald, 137

Klehini River, 94, 216, 233

Klemetson, Ole (Haetta), 184

Klondike, 15, 17, 37, 51, 52, 101, 107, 134, 197, 214, 218, 220

Klukwan (village), 93, 99–101, 104, 119, 215, 216

Knox, William S. (representative, Massachusetts), 18

Kobuk River, 227

Kodiak, Kodiak Island, 8, 123

Koriak people (Siberia), 227

Kotzebue Sound, 221, 227

Krogh, Ole, 111, 116, 121, 137, 142, 177, 181, 205

L

Lacey, John F. (representative, Iowa), 18

Lakeview Cemetery (Seattle), 214

Lakselv, Peder Johannesen, 97

Lakselv, Samuel Josefsen (*see* Josefsen, Samuel), 97

Lane, Charles D. (Wild Goose Mining Co.), 201, 205, 206, 228, 229

Langfjord, 4, 63, 66, 237

Lantis, Margaret, 234

Lapps (Saami), 28, 34, 37, 38, 54, 55, 60, 65–67, 69, 72, 74, 75, 77, 80–82, 85, 93, 108–110, 112–115, 119, 120, 130, 131, 139, 143, 145, 148, 149, 154, 155, 164, 165, 167, 170, 173–175, 177, 179, 180, 185, 197, 201, 202, 205, 211, 212, 214, 217, 218

Larsen, Frederik, 131, 160, 161, 179, 219, 223

Larsen, Lauritz, 111, 181

Larsen, Per, 189

Laub, Captain (steamship *Thingvalla*), 212

*Laurada* (steamship), 205, 207, 229, 230

Law and Order League, 226

Leinan, Otto, 128, 142, 161, 177, 180, 181, 188, 224

*Leo* (schooner), 50

Libby, D. B. (gold miner, Council City), 219

Lichens (*Cladonia* spp., *Cetraria* spp.) (reindeer moss), 3, 6, 26, 30, 32, 33, 40, 41, 47, 50, 56, 57, 66, 86, 87, 89, 90, 98, 101–104, 212, 235

Lie, August, 198

Lieber, G. Norman (judge advocate gen., U.S. Army), 217

Lindblom, Eric O. (gold miner), 156, 157, 191–193,198, 203, 204, 223, 224

Lindeberg, Jafet, 75, 80, 94, 97, 99, 101, 104, 105, 109, 114–116, 119, 121, 131, 139, 145, 151, 156–161, 186, 188, 190–193, 196, 198–202, 205, 207, 209, 218, 222–224, 227, 228

London, 19, 22–26, 29, 30, 38, 42, 45, 213

Long, John D. (secretary of navy), 51

Lopp, William T., 14, 51, 235

Loppen (Loppa) Island, 38, 41, 66, 210

Losvar, Johan, 74, 93, 96, 180, 181

*Louise J. Kenney* (schooner), 3, 49, 118, 120, 130, 140, 150, 218, 219, 221

*Lucania* (steamship), 21, 22

Lumijärvi, Hannah (Johannah) (Mrs. Carl Sacarisen), 237, 238

Lyng, R. T. (mgr., Alaska Commerical Co., and postmaster, St. Michael), 221, 227

M

McBride, George W. (senator, Oregon), 16

McFarland, Mrs. A. R. (teacher, Fort Wrangell), 49

McKinley, President William, 12, 13, 15, 16, 19, 45, 51, 229

Mail service, in Alaska, 2, 52, 150, 161, 162, 215, 217, 220–222, 235

*Maine* (battleship), 217

Malimiut, dialect (language), 50

Manila (Philippine Islands), 107

*Manitoban* (steamship), 4, 28, 29, 32–38, 40, 41, 47, 48, 57, 63, 65, 69, 70, 72, 73, 76, 210–215

Marshall, John (chief justice, U.S. Supreme Ct.), 228

Martensdatter, Inger Maria, 214

Martin (Iñupiaq, herder), 136

Masi (town), 34

Meiklejohn G. D. (asst. secretary of war), 29, 44, 46, 55, 58, 211, 212, 217, 218

Melsing, Louis F. (gold miner, Council City), 219, 224

Merriam, H. C. (brigadier gen.; commander, Dept. of the Columbia), 52, 213, 215, 217

Metlakatla (village, S.E. Alaska), 230

*Mildred* (river steamer), 218

Miles, Nelson (gen., U.S. Army), 212

Miller, James M. (physician, U.S. Army), 207, 209, 229

Milwaukee (town), 77

*Milwaukie* (river steamer), 218

Mining claims, under dispute, 170, 171, 181, 193–197, 201, 202, 225–229

*Minneapolis* (river steamer), 122

Mocatta, Jansen and Company (brokerage, foreign produce), 24, 25, 29, 30, 32, 33, 38, 41, 42, 44–46

Mordaunt, A. P. (gold miner, Council City), 219, 226, 227

Morgan, J. D. (gold miner), 192, 198

Moses Point (village), 185

Mosjøen (town), 30, 33, 40

Muir, John, 50

Murray, Keith, 236, 237

N

Nafta (medical compound), 149, 222

Nakkila, Bereth Anna (Berit), 187, 207

Nakkila, Isak Salomonsen, 169, 192, 197, 231

Nakkila, Mikkel (Näkkälä), 131, 135, 142, 181, 187, 192, 207, 224, 227

Nanaimo (British Columbia), 13

*Navarro* (steam schooner), 120, 121, 123, 124, 218

Navy, secretary of the, 51

Nelson, E. W. (ethnologist), 222

New York, Brooklyn Bridge, 74, 212

New York, city and harbor, 44, 69, 71–75, 211, 212

New York, Municipal Building, 74, 212

*New York Sun*, 60, 212

*New York Times*, 13, 212

Nilima, Alfred, 181

Nilsen, Klemet, 146, 165, 167, 168, 223

Nome (*see* Anvil City), 177, 221, 225, 226, 227, 234

Nome, Cape, 11, 151, 156–162, 164, 169–173, 176, 177, 179–181, 184, 187, 188, 193, 194, 197, 199, 200, 207, 209, 222–225, 227, 228, 230

*Nome Gold Digger* (newspaper), 230

Nome Mining and Development Company (Kjellmann et al., owners), 227

Nome River, 177, 188–190, 201, 205, 206

North American Commercial Company, 220

North American Transportation and Trading Company, 52, 126, 160, 169, 170, 188, 195, 197, 201, 221, 224, 225, 231

Norton Sound (Norton Bay), 3, 150, 183, 184, 235

Norway, 4, 19, 22, 25–30, 32–34, 36, 38, 40, 42–46, 63, 74, 75, 88, 108, 113–116, 126, 130, 134, 140, 173, 179, 204, 209, 211, 214, 215, 217, 220, 223, 226, 228, 235, 236

Norway, Meterological Institute of, 26

Norway, Royal Military Service, 4

Nulato (village), 150, 155, 159, 165, 223

Nulato Hills, 50

Nunivak Island, 235

O

Olsen, Ole, 172

Omekejook, Alice (teacher, Unalakleet), 49, 166, 219, 223

*Oregon Journal* (newspaper), 236

O'Reilly vs. Campbell (Sup. Ct., 1886), 228

Organic Act, Alaska (1884, S. B. 153; Sen. Benjamin Harrison, Indiana, sponsor), 6, 224

Osborn Creek (Seward Peninsula), 189, 190

Oscar II, King (Norway), 116

*Ottawa* (steamship), (*see Manitoban*), 28

P

Pacific Coast Steamship Company, 220

Pacific Steam Whaling Company, 220

Paulsen, A. (merchant, Norway), 34, 59

Paulsen, Olai, 119, 146, 148, 156–158, 163, 165, 171, 174, 176, 192, 205, 207, 227

Pauncefote, Sir Julian (ambassador to U.S. from Great Britain), 45, 46

Pennsylvania Railroad Co., 211, 212

Permafrost, Seward Peninsula, 225

Pioneer Mining Co. (Cape Nome), 193, 195, 198, 206

Pittsburgh, 76

Point Barrow, Barrow, 221, 233

Point Hope, 221

Porsanger, Per, 111

Port Clarence (*see* Teller Reindeer Station), 11, 50, 130, 135, 137, 151, 179, 181, 199, 219, 220, 224, 225, 227

*Portland* (steamship) (formerly *Haytien Republic*), 51, 205, 207–209, 231

Port Townsend, 83, 107–110, 114, 116, 119, 121, 122, 130, 139, 216, 218, 227

*Portus B. Weare* (river steamer), 11, 12, 51

Postal service, Alaska (*see* Mail service)

Post Office Department, U.S., 220, 221

Presbyterian Church, 2, 8, 10, 49, 56, 229, 234

Presbyterian Mission School, St. Lawrence Island, 13, 219

Presbyterian missions, 7, 49

President, U.S., 12, 13, 15, 16, 19, 24, 45, 51, 220, 224, 234

Pribilof Islands, 8, 50, 230, 235

Price, Gabriel W. (gold miner), 157, 191, 193, 195, 201, 223, 225, 227

Price, R. L. (gold miner), 192, 225, 227

Protestant Episcopal Mission, 233

Prydz, F. (judge), 46

Ptarmigan hunting, 169

Pulk, Ole Johannessen, 231

Pyramid Harbor, 95, 215

## Q

Quarantine Act of 1893, 211

Quarantine: Camp Low Federal Quarantine Station, 211

Quartermaster, U.S. Army, 211, 215, 217

*Queen* (steamship), 216

Queenstown (Ireland), 43

Qvigstad, K. (judge), 46

## R

Rafferty, J. J. (special agent and scout, Dept. of War), 211, 212, 214

Randall, George M. (col., 8th Infantry), 15, 51

Rapp, Ole, 94, 118, 181, 204

Rauna, Johannes, 101, 103, 116, 121, 137, 181

Ray, Patrick Henry (capt., U.S. Army), 15, 51, 52

Redmyer, Hedley E., 81, 100–102, 104, 119, 213, 216, 218, 233

Reed, Thomas B. (speaker, House of Representatives), 19

Reindeer, disease, 25, 56, 167, 214

Reindeer Expedition, 25, 64, 73, 76, 80, 150, 199, 218, 219, 222

Reindeer Experiment Station (Biol. Survey, U.S. Dept. of Agriculture), 235

Reindeer industry (Alaska), 2, 6, 49, 217, 234

Reindeer moss, 3, 6, 26, 32, 33, 40, 41, 47, 50, 56, 57, 66, 86, 87, 89, 90, 98, 101–104, 212, 235

Revenue Marine Cutter Service, U.S., 8–11, 13, 14, 51, 198, 199, 206, 221, 230

Richardson, Percival C. (postal service), 139, 150, 221

Richardson, Wilds P. (1st lt., U.S. Army), 15, 51, 52

Rist, Johan, 34, 110–112

Rist, Per Aslaksen, 36

Rivertz, Peter (Lensmann), 40–46

*Roanoke* (steamship), 196, 205, 207, 229

Robins, Raymond (chief of staff, Rev. Wirt), 230

Robinson, William W. (capt., Quartermaster Dept., U.S. Army), 214

Roosevelt, President Theodore, 234

Røros (town), 26, 56

Rucker, L. H. (maj., 4th Cavalry, U.S. Army), 214

Russia, 7, 8, 11, 54, 108, 127, 219

## S

Saami (Lapps), 28, 34, 37, 38, 54, 55, 60, 65–67, 69, 72, 74, 75, 77, 80–82, 85, 93, 108–110, 112–115, 119, 120, 130, 131, 139, 143, 145, 148, 149, 154, 155, 164, 165, 167, 170, 172–177, 179, 180, 185, 197, 201, 202, 205, 211, 212, 214, 217, 218

Sacarisen, Hannah Lumijärvi (Sakariassen), 237, 238

*Sadie* (steamship), 200

St. George Island (Pribilof Islands), 8, 230

St. Lawrence Bay (Siberia), 220

St. Lawrence Island, 13, 14, 116, 120, 121, 128, 198, 219, 227, 235

St. Leger, A. (purchasing agent), 220

St. Michael, Fort (U.S. Army), 15, 51, 127, 223

St. Michael (village), 1, 11–15, 50–52, 109, 113–115, 121, 124–129, 137, 139, 140, 142, 146, 150, 151, 158–161, 163–165, 168–170, 172, 174, 177, 179, 181, 192–196, 198, 200–202, 205, 218–221, 223–225, 227, 231

St. Paul Island (Pribilof Islands), 8

Sakariassen (Sacarisen), Carl Johan, 4–6, 36, 60, 63, 80, 88, 179, 189, 190, 192, 208, 209, 211, 213–215, 217–225, 227, 230, 231, 236–238

Sampson Company (Anvil City), 197

Samuelsen, Hans, 104, 105

Sandy Hook, New Jersey (Camp Low Fed. Quarantine Sta.), 211

San Francisco, 12, 14, 205–209, 220

*San Francisco Chronicle*, 13

Sara, Mattis Nilsen, 214

Sara, Nils Person, 214

Sayers, Joseph D. (representative and gov., Texas), 18, 53

Schiander, Arndt (attorney), 41, 43, 44

Schjodt and Foss (attorneys), 41

School, first U.S., established in District of Alaska, 49

Schrader, Frank C., 227

Scurvy, 151, 155, 162, 170, 173, 176, 194, 222, 227

*Scythia* (steamship), 217

*Sea Lion* (steam tug), 83, 214

Seattle, 3, 12, 14, 79, 80, 82, 84, 114, 118, 120, 122, 124, 146, 157, 205, 209, 211, 212, 214, 217, 219, 221, 229, 230

*Seattle Daily Times*, 215, 217

Seattle Hardware Co., 218

*Seattle Post-Intelligencer*, 213, 216, 218, 220, 230

Seattle-Skagway Steamship Co., 218

Seim, Conrad (purchasing agent), 220

Seligman Brothers, Bank (London), 45, 213

*Seminole* (steamship), 86, 94, 214, 215

Seppala, Leonhard, 237

Seward Peninsula, 225, 226

Seward, William H., 24, 225

Shaktoolik (village), 182

Shephard, Lennox B. (U.S. commissioner and judge, St. Michael), 224–226

Sherman, John (secretary of state), 21, 55

Siberia, 2, 8, 10, 11, 116, 121, 131, 132, 135, 137, 198, 220, 227

Sickness, in people, 9, 72, 96, 114, 146, 151, 155, 162, 168, 170, 173, 176, 177, 194, 205, 206, 222, 227

Simpson, Jerry (representative, Kansas), 18

Sinrock River, 194, 225

Siri, Hans Andersen, 233

Siri, Per Nilsen, and herding dog, 233

Sitka, 154, 198, 202

Sivertsen, S. N. (brokerage, commodities), 30, 33, 40–42, 45, 46

Sivuqaq (Gambell), St. Lawrence Island, 219

Skagway, 91, 107, 108, 134, 214, 215

*Skandinaven* (newspaper), 216

Sledge Island, Bering Sea, 227

Smith, J. U. (U.S. commissioner), 216

Snake River (Seward Peninsula), 156–158, 171, 177, 188, 189, 194, 196, 200, 206, 225

Somby, Aslak L., 36

Southward, Arthur E. (gold miner, physician), 150, 162, 163, 168, 169, 171, 172, 207, 223

Spanish-American War, 107, 215, 217, 224

Spaulding, Oliver L., Jr. (2d lt., 3d Artillery Regiment, U.S. Army), 224, 226

Spooner, John C. (U.S. Senate, Wisconsin), 226

Spring, A. (mining engineer), 150, 161, 162, 171, 179, 180, 188

State, secretary of, 11, 21, 45

State, U.S. Dept. of, 45, 46

Statue of Liberty, 74

Stefansen, Lauritz, 74, 118, 181, 204, 207

Stenersen, Jonas (judge), 46

Stensfjeld, Ole, 67, 174, 176

Stewart, William M. (senator, Nevada), 228

Suhr, Carl, 34, 59, 186–189

Swedish Evangelical Mission Covenant, 2, 49, 219, 222, 223, 225

Swedish Mission Church, Golovnin Bay, 235

T

Tanana (village) (Weare), 150, 221, 223, 233

Tanana River, 11, 221, 233

Teller, Henry M. (senator, Colorado), 50, 51, 228

Teller Reindeer Station (Port Clarence; Teller village), 11, 50, 130, 135, 137, 151, 179, 181, 199, 219, 220, 224, 225, 227

Thanksgiving Day, Eaton Station, 155–159, 222

*Thingvalla* (steamship), 73, 212

Tlingit people, 87, 88, 214, 216

Togo (L. Seppala's lead dog), 236

Tonseth, E. A. (brokerage), 26, 32, 56

Tornensis, Isak, 174, 176

Tornensis, J. (Johan Isaksen), 66, 111, 136, 157, 184, 209

Tornensis, Johan Sp., 131, 135, 136, 142, 181, 192, 219, 223, 224, 227

Treasury, Dept. of, 13, 20, 50

Treasury, secretary of, 8, 11, 13, 51, 212

Trickett, Arthur (veterinarian), 212

Tromsø, 26, 33, 38, 41

Trondheim, 24–28, 30–33, 38, 40, 41, 47, 212

*Trondhjems Adresseavis* (newspaper), 37

Tununak Village (Tanunak), 14

Tuttle, Francis (U.S. Revenue Marine Service), 13, 221

Typhoid fever (Nome), 205–208, 229

Tyson, James (fishing fleet), 221

U

Unalakleet (village), 1–3, 6, 8, 49, 50, 121, 128–130, 133, 134, 136, 137, 146, 150, 151, 157–159, 163–169, 174–177, 179–182, 185, 194, 199, 219–221, 223, 224, 227, 235

Unalakleet River, 2–4, 6, 150, 164, 220, 222, 223, 235

Unalaska (Dutch Harbor), 125, 208, 220, 230

Updegraff, Harlan (Bur. of Education), 234

Uparazuck, Konstantin (Constantin), 157, 192, 223, 225

U.S. Marine Hospital Serv., 211

V

Victoria, B.C. (Vancouver Island), 108, 209, 217

*Victoria Daily Colonist* (newspaper), 51

Vocational schools, Alaska, 2, 10, 49

Vorren, Ørnulv, 213, 219

W

Walker, Edgar S. (capt., CO, 8th Infantry Regiment, U.S. Army), 161, 170, 202, 219, 223, 225, 226

Walrus, 9

War, Department of, 2, 12, 21, 23, 26, 27, 29, 35–38, 40–42, 44–49, 51, 211–215, 217, 219, 222

War, secretary of, 15, 16, 19, 20, 21, 23, 24, 27–29, 42–45, 47, 48, 51–55, 57, 58, 212–214

Washington and Alaska Steamship Co., 217

Watterson, John (gold miner), 224

Weare, Ely (No. Amer. Transp. and Trading Co.), 52

Weare (village) (Tanana), 150, 221, 223, 233

Wedding ceremony, Saami (Eaton Station), 174, 175

Wells Fargo Company, 220

West Coast Steam Navigation Company, 218, 220

Weston, E. W. (lt. col., chief commissary officer), 211

Whaling fleet, arctic, 10, 12, 13

Whaling vessels, 9

Wheeler, Charles K. (representative, Kentucky), 18

White, Frank M. (captain, *Laurada*), 230

Wiig, Rolf, 116, 120, 143, 161, 181, 203, 204

*Wilbur Crimmon* (river steamer), 218

Wild Goose Mining Company, 228

Wilson, James (secretary of agriculture), 11

Wilson, J. L. (gold miner), 226, 227

Windon, William (secretary of the treas.), 11

Wines, Arthur Frederick (U.S. Census Bur.), 225, 228

Wirt, Loyal L. (missionary, Presbyterian Church), 229, 230

*Wisconsin* (dredger barge), 218

Wolfe, Forbes Hall (surgeon, *Manitoban*), 211

Women's Christian Temperance Union, 50

Wood, Leonard (major gen., U.S. Army), 57

Woodland Park, Seattle, 80, 87, 215

*World* (newspaper, New York), 60

Wrangell (Fort Wrangell), 49, 108, 217

Wulff (attorney), 41, 43

## Y

Yeatman, Richard T. (capt., 14th Infantry), 217

Yukon District (Territory), Canada, 1, 2, 12, 13, 16, 19, 21, 25, 30, 47, 54, 139, 142, 155,, 180, 219, 220, 226

Yukon Relief Expedition (mission), 2, 4, 6, 17, 25, 27, 29, 33, 37, 40, 42, 44, 46, 47, 211–214, 217–219, 222, 227, 228, 233, 236

Yukon River, 2, 8, 11, 14, 15, 52, 94, 104, 139, 150, 169, 218, 220, 221, 223, 227, 233

Yup'ik names, spelling, 224

## Z

Zapadni Bay (St. George I., Pribilof Islands), 230

# About the Editors

...who add their father's journal to the record of events in northern regions during the late nineteenth century.

From 1950 to 1975, V. R. (Reggie) Rausch was involved in biomedical investigations with the United States Public Health Service in Alaska. She has continued faunal studies there, and in recent years has worked also in arctic and subarctic regions of Canada and Siberia, and in northern China. She is an affiliate in mammalogy at the Burke Memorial Washington State Museum at the University of Washington. Her other interests include history and languages.

D. L. Baldwin was associated for many years in management of her family's business in Eugene, Oregon. In concern for the biosphere, she has long supported efforts to promote conservation and to preserve natural areas, especially in western North America. There she has traveled widely, often hiking in the high Cascades of Oregon and in the arid regions adjoining and to the south. Dana Baldwin has a strong interest as well in the language and culture of Norway.

the authors are donating
the royalties from this publication to
The Nature Conservancy